Handbook of CCD Astronomy

Charge-Coupled Devices (CCDs) are the state-of-the-art detector in many areas of observational science. Astronomy, physics, chemistry, medical imaging, and remote sensing are just a few of the fields of study in which they play a crucial role. This handbook provides a concise and accessible reference on all practical aspects of using CCDs.

Starting with the electronic workings of these marvels of modern science, this handbook discusses the basic characteristics of CCDs and then gives methods and examples of how to determine these values. While the focus of the book is on the use of CCDs in professional observational astronomy, advanced amateur astronomers and researchers in physics, chemistry, medical imaging, and remote sensing will also find it very valuable. The core of the book focuses on the detection of optical light, but due coverage is given to other wavelengths, particularly high-energy astronomy. Tables of useful and hard-to-find data and key practical equations round the book off and ensure that it provides an ideal introduction to the practical use of CCDs for graduate students, as well as a handy reference for more experienced researchers.

Dr. Steve Howell was raised in the eastern United States. Starting his research career with laser physics work at General Motors Laboratories, he eventually moved west to work at Kitt Peak National Observatory where he supported software and hardware development for CCD instrumentation. Since that time, Howell has worked at Goddard Space Flight Center as a member of the International Halley Watch, developing image processing software for use by NASA and the Shuttle programs, was a team member for the Galileo Spacecraft imager, worked at the Planetary Science Institute as a staff scientist, and was a professor of astrophysics at the University of Wyoming. Currently, Howell is an adjunct faculty member of the Physics and Astronomy Department at Wyoming and Head of the Astrophysics Group of the Planetary Science Institute in Tucson, Arizona.

D0709033

Cambridge Observing Handbooks for Research Astronomers

Today's professional astronomers must be able to adapt to use telescopes and interpret data at all wavelengths. This series is designed to provide them with a collection of concise, self-contained handbooks which covers the basic principles peculiar to observing in a particular spectral region, or to using a special technique or type of instrument. The books can be used as an introduction to the subject and as a handy reference for use at the telescope, or in the office. They also promote an understanding of other disciplines in astronomy and a modern, multi-wavelength, multi-technique approach to research. Although aimed primarily at graduate students and researchers, many titles in the series are of interest to keen amateurs and undergraduate students.

Series editors
Professor Richard Ellis, Institute of Astronomy, *University of Cambridge*
Professor John Huchra, Center for Astrophysics, *Smithsonian Astrophysical Observatory*
Professor Steve Kahn, Department of Physics, *Columbia University*, New York
Professor George Rieke, Steward Observatory, *University of Arizona*, Tucson
Dr Peter B. Stetson, Herzberg Institute of Astrophysics, *Dominion Astrophysical Observatory*, Victoria, British Columbia

HANDBOOK OF CCD
ASTRONOMY

STEVE B. HOWELL

*ASTROPHYSICS GROUP, PLANETARY SCIENCE INSTITUTE, TUCSON,
AND DEPARTMENT OF PHYSICS AND ASTRONOMY,
UNIVERSITY OF WYOMING*

CAMBRIDGE
UNIVERSITY PRESS

PUBLISHED BY THE PRESS SYNDICATE OF THE UNIVERSITY OF CAMBRIDGE
The Pitt Building, Trumpington Street, Cambridge, United Kingdom

CAMBRIDGE UNIVERSITY PRESS
The Edinburgh Building, Cambridge CB2 2RU, UK
40 West 20th Street, New York, NY 10011-4211, USA
10 Stamford Road, Oakleigh, Melbourne 3166, Australia
Ruiz de Alarcón 13, 28014 Madrid, Spain
Dock House, The Waterfront, Cape Town 8001, South Africa

http://www.cambridge.org

© Cambridge University Press 2000

First Published 2000
Reprinted 2001

Printed in the United States of America

Typeface Computer Modern 10/13 pt.   *System* LATEX [TB]

A catalog record for this book is available from the British Library.

Library of Congress Cataloging in Publication Data
ISBN 0 521 64058 X hardback
ISBN 0 521 64834 3 paperback

Contents

Preface

We are all aware of the amazing astronomical images produced with telescopes these days, particularly those displayed as color representations and shown off on websites and in magazines. For those of us who are observers, we deal with our own amazing images produced during each observing run. Just as spectacular are photometric, astrometric, and spectroscopic results generally receiving less fanfare but often of more astrophysical interest. What all of these results have in common is the fact that behind every good optical image lies a good charge-coupled device.

Charge-coupled devices, or CCDs as we know them, are involved in many aspects of everyday life. Examples include video cameras for home use and those set up to automatically trap speeders on British highways, hospital x-ray imagers and high speed oscilloscopes, and digital cameras used as quality control monitors. This book discusses these remarkable semiconductor devices and their many applications in modern day astronomy.

Written as an introduction to CCDs for observers using professional or off-the-shelf CCD cameras as well as a reference guide, this volume is aimed at students, novice users, and all the rest of us who wish to learn more of the details of how a CCD operates. Topics include the various types of CCD; the process of data taking and reduction; photometric, astrometric, and spectroscopic methods; and CCD applications outside of the optical bandpass. The level of presentation is aimed not only at college or professional level readers but also at a more general audience including the ever growing number of highly trained and motivated amateurs and other professionals in technical areas in which CCDs play a role.

Chapters 2 and 3 contain all the fundamental information on CCD operation and characteristics while each remaining chapter can be mastered individually. In a book of this length, many aspects must be treated briefly. However, I have made an effort to provide self-contained detail of the important aspects of CCDs while including numerous references to the detailed professional literature for those desiring a deeper understanding. Additionally, throughout the book, examples related to common observational occurrences as well as footnotes discussing interesting but not mainstream topics are included. Appendices list other reference works of interest, CCD manufacturers, numerous website addresses of interest, and a brief introduction to image displays.

This book started with an idea for a new series of handbooks, including a volume on CCDs. I am happy to thank the editor, Adam Black, and Peter Stetson for allowing me to be involved in this series and write this book. The folks at Cambridge University Press, particularly Adam, have been very helpful, dealing with my many questions during the writing process. Michie Shaw and the staff at TechBooks have helped greatly in the final steps of production. I would like to thank the anonymous readers of an early draft of this book for their comments and for pointing out some important areas and results I had overlooked. Two readers, Peter Stetson and John Huchra, suggested coverage of material that has led to the inclusion of additional topics in the final version. A number of colleagues have provided information, graphs, references, and support during the writing of this book, all of which I appreciate. I thank my former and current students and postdocs for keeping me on my toes regarding many of the topics herein. Chris Sabbey kindly provided Figure 6.8, and Figures 1.1 and 4.6 were taken by Simon Tulloch and provided by Derek Ives, both of the UK Astronomy Technology Centre.

I would like to acknowledge and thank my parents, Cecil and Barbara, for allowing me to be "scientific" during my childhood. My experimentation, often at the expense of household items, was not always successful but was never discouraged. These experiences deeply planted the seed of scientific fascination in my being. Appreciation is also passed along to my brother Terry, for the many hours we spent together exploring the world around us. Particularly noteworthy were the times we spent watching, analysing, and laughing at "B" sci-fi movies.

During the writing of this volume on CCDs, many opportunities were missed related to spending time with my son Graham and my wife and friend Mary. Both were always supportive of the effort, encouraged its completion, and have accumulated many IOUs, which I will now have the pleasure of paying off. I appreciate their unfailing love.

I have had fun writing this book and learning even more about CCDs, almost as much fun as I have when I observe with them. I hope that you the reader will find this work of interest as well and enjoy paging through it often. Astronomy has always fascinated humans and if this treatise allows you to obtain a better knowledge of CCDs and with it even more fascination with the universe around us, it will have been a success.

"Since The Beginning Of Time, The Universe Has Called To Awaken Each Of Us. To Understand The Universe Is To Understand Ourselves."

1

Introduction

Silicon. This semiconductor material certainly has large implications on our life. Its uses are many, including silicon oil lubricants, implants to change our body's outward appearance, electric circuitry of all kinds, nonstick frying pans, and of course, charge-coupled devices.

Charge-coupled devices (CCDs) and their use in astronomy will be the topic of this book. We will not discuss the use of CCDs in commercial video cameras or many other industrial and scientific applications. As we will see, there are four main methods of employing CCD imagers in astronomical work: imaging, astrometry, photometry, and spectroscopy. Each of these topics will be discussed in turn. Since the intrinsic physical properties of silicon, and thus CCDs, are most useful at optical wavelengths (about 3,000 to 10,000 Å), the majority of our discussion will be concerned with visible light applications. Additional specialty or lesser used techniques and CCD applications outside the optical bandwidth will be mentioned only briefly.

Applications of infrared (IR) arrays – semiconductor devices with some similar properties to CCDs – while important to modern astronomy, will receive small mention here. A complete treatment of IR arrays is given by Ian Glass in his companion book, *Handbook of Infrared Astronomy*, the first book in this series.

Appendix A provides the reader with a detailed list of other books and major works devoted to CCD information. This appendix does not contain the large body of journal articles in existence; these will be selectively referenced throughout the text. Appendix B provides a list of present day CCD manufacturers that produce both amateur and professional grade CCDs and CCD cameras. Finally, Appendix C discusses some basic principles of image display devices. While not directly related to CCDs, computer displays are the medium by which we view and examine the information collected by CCDs. Proper interpretation of the CCD image is only possible if one understands how it is displayed.

1.1 Nomenclature

CCDs are often listed and named by an apparently strange convention. This small section aims to demystify these odd sounding names. CCDs come is various sizes and shapes and are manufactured by a number of companies (see Appendix B).

Figure 1.1 illustrates a few typical CCDs. Present day CCDs generally come is sizes ranging from 512 by 512 "picture elements" or pixels to arrays as large as 4,096 by 4,096 pixels. Often these size designations are written as 512×512 or 512^2. CCDs are also available as rectangular devices of unequal length and width and with nonsquare pixels. For example, CCDs of size 1024×3072 pixels are produced for spectroscopic applications. We will see in Chapter 2 that each pixel acts as an electrically isolated portion of the silicon array and is capable of incoming photon collection, storage of the produced photoelectrons, and readout from the CCD array to an associated computer as a digital number.

The names or designations of CCDs are usually a combination of the company name with the CCD size. RCA512, $4K \times 2K$ EEV, or SITe2048 are examples. Instrumentation at observatories often include a CCD as the detector and are specialized to perform a task such as imaging or spectroscopy. Observatories designate these instruments with a name that may or may not include information about the associated CCD detector. The Royal Greenwich Observatory (RGO) on La Palma has the FOS#1A (a 512×1024 Loral CCD used in their Faint Object Spectrograph), and the Tek 2K CCD of the National Optical Astronomy Observatories (NOAO) is a 2048×2048 pixel array used in their 0.9-m telescope imaging camera. Observatories keep lists of each of their CCDs and instruments with associated information about the CCD specifications. For examples of such information, check out the observatory websites listed in Appendix B.

1.2 Why Use CCDs?

Most astronomical detectors in use today at professional observatories, as well as with many amateur telescopes, are CCDs. This fact alone gives the reader an impression that there must be something very special

Fig. 1.1. A selection of recent CCDs that have been placed in operation at the Royal Greenwich Observatory. Clockwise from bottom left they are: SITe-002 (2048 × 4096), Loral 2k3eb (2048 × 2048), EEV CCD42-80 (2048 × 4096), SITe-424 (2048 × 2048), GEC P8603 (385 × 578), EEV 15-11 (1024 × 256), TeK1024 (1024 × 1024), Loral 512FT (512 × 1024), EEV-05-30 (1242 × 1152), EEV CCD42-10 (2048 × 512), Loral-64 (64 × 64), and EEV CCD39-01 (80 × 80).

or useful about CCDs; otherwise why all the fuss? CCDs have revolu-
tionized modern astronomy. They will take their place in astronomical
history along with other important discoveries such as the telescope,
photographic plates, prisms, and spectroscopy. The contribution to our
knowledge of the heavens brought about by CCDs is astounding, even
more so when one considers that they have been in use for only about
twenty-five years.

First introduced as electronic analogues to magnetic bubble memory (Boyle & Smith, 1970; Amelio, Tompsett, & Smith, 1970) at Bell
labs, CCDs provided their first astronomical image in 1975 when scien-
tists from the Jet Propulsion Laboratory imaged the planet Uranus at
a wavelength of 8,900 Å (Janesick & Blouke, 1987). This observation
used the University of Arizona 61-inch telescope atop Mt. Lemmon and
a prototype version of a CCD made by Texas Instruments Corporation
as part of a development project for NASA spacecraft missions.

During the past quarter century, tremendous progress has been made
in the manufacturing process and therefore in the properties of the CCD
itself. These improvements have allowed much lower noise properties for
CCDs, thereby increasing their overall astronomical usage. We outline
below two of the important reasons why CCDs are considered as essentially the perfect imaging device. Details of the manufacturing techniques
and properties of CCDs will be presented in Chapters 2 and 3.

1.2.1 Noise Properties

The usefulness of a detector is very often determined by the amount of
inherent noise within the device itself. We shall see in Chapter 3 how
the noise properties of a CCD are determined, but, suffice it to say
here, modern CCDs are almost noise free. The original line of photosen-
sitive electronic array detectors, such as television-type imaging detectors, vidicons, silicon intensified targets, and image-dissector scanners,
all had very high noise properties. For comparison, silicon intensified target imagers (SITs) had a noise level upon readout of 800 electrons per
picture element. Some very good systems of this type could be produced
with read noise values of only 200 electrons (Eccles, Sim, & Tritton,
1983). The first CCDs had readout noise levels similar to this latter
value, while modern CCDs have noise values of fifteen to only a few
electrons per pixel. The large noise levels present in early array detectors not only limited the signal-to-noise ratio obtainable for a given

measurement, they also severely limited the total dynamic range available to the camera.

1.2.2 Quantum Efficiency and Bandpass

Quantum efficiency (QE) is the term used to report on the ability of a detector to turn incoming photons into useful output. Generally, it is defined as the ratio of incoming photons to those actually detected or stored in the device. A QE of 100% would be an ideal detector with every incoming photon detected and accounted for in the output. Bandpass is a term that means the total spectral range for which a detector is sensitive to the incoming photons. Our eyes, for example, have a very limited bandpass covering only about 2,000 Å of the optical spectral range, from near 4,500 to 6,500 Å.

One of the great advantages of CCDs compared with earlier detectors is their ability to convert a large percentage of incoming photons into photoelectrons. Photographic plates had an intrinsic quantum efficiency of only about 2% (Kodak IIIaJ's reached 3%), with "hypersensitized" plates (plates treated to a black-magic process involving heating and exposure to various gases) reaching claimed QEs as high as 10%. Because photographic emulsions were originally more sensitive to UV and blue light, numerous dyes and coatings were developed to both extend their bandpass coverage and allow detection of yellow to red optical photons.

Early solid-state imaging devices and intensified target devices could reach quantum efficiencies as high as 20–40%. These devices relied on television-type tube technology and electron beam scanning for readout of detected photons. Since they used silicon (or similar type materials) as the detector material, their useful spectral bandpass was similar to that of modern CCDs. However, besides the relatively low QE, these early electronic detectors had other drawbacks. For example, image tubes needed high voltage input for electron acceleration and the observed two-dimensional scene was not easily or consistently divided into well-determined x, y positional output (Walker, 1987). Changes in the voltage levels, collected charge, and telescope position resulted in electric and magnetic field variations leading to positional and flux measurement uncertainties.

Even the earliest CCDs (those manufactured by Fairchild or GEC) easily reached peak QEs of 40%. Typical CCD QE curves today not only peak near 90% but are 60% or more quantum efficient over two thirds of their total spectral range. The bandpass available in a modern CCD (with a QE of 10% or more) is about 3,000–10,500 Å. Coatings and phosphors deposited on the CCD surface or the use of some form of pre-processor device can extend the bandpass sensitivity limits or increase the QE in specific wavelength ranges (see Chapter 2).

This volume of the Cambridge Observing Handbooks for Research Astronomers will explore the world of CCDs from their inner workings to numerous applications in observational astronomy. Appendices are included to provide ancillary information related to the main text. The chapters will make little assumption as to the readers previous knowledge on the subject, each attempting to be somewhat self-contained. Chapters 4, 5, and 6 deal directly with astronomical applications while Chapters 2 and 3 are of general interest to those wanting an overall understanding of CCDs as detectors. Chapter 7 discusses the use of CCDs at nonoptical wavelengths. In a short treatise such as this, coverage of numerous details and nuances is not possible; thus a detailed reference list to general texts or collections of articles on CCDs is provided in Appendix A. For those wishing to explore a subject at a deeper level, pertinent research articles are provided throughout the text itself.

2

CCD Manufacturing and Operation

Before we begin our discussion of the physical and intrinsic characteristics of charge-coupled devices (Chapter 3), we want to spend a brief
moment looking into how CCDs are manufactured and some of the basic,
important properties of their electrical operation.

The method of storage and information retrieval within a CCD is
dependent on the containment and manipulation of electrons (negative
charge) and holes (positive charge) produced within the device when exposed to light. The produced photoelectrons are stored in the depletion
region of a metal insulator semiconductor (MIS) capacitor, and CCD
arrays simply consist of many of these capacitors placed in close proxim-
ity. Voltages, which are static during collection, are manipulated during
readout in such as way as to cause the stored charges to flow from one
capacitor to another, providing the reason for the name of these devices.
These charge packets, one for each pixel, are passed through readout
electronics that detect and measure each charge in a serial fashion. An
estimate of the numerical value of each packet is sent to the next step
in this process, which takes the input analog signal and assigns a digital
number to be output and stored in computer memory.

Thus, originally designed as a memory storage device, CCDs have
swept the market as replacements for video tubes of all kinds owing
to their many advantages in weight, power consumption, noise characteristics, linearity, spectral response, and others. We now further explore some of the details glossed over in the above paragraph to provide
the reader with a basic knowledge of the tortuous path that the detected photon energy takes from collection to storage. The design of
CCD electronics, semiconductor technology, and detailed manufacturing methods are far beyond the level or space constraints of this book.
For further information the reader is referred to the excellent discussion

in Janesick & Elliott (1992) and the other technical presentations listed in Appendix A.

2.1 CCD Operation

The simplest and very understandable analogy for the operation of a CCD is also one that has been used numerous times for this purpose (Janesick & Blouke, 1987). This is the "water bucket" idea in which buckets represent pixels on the CCD array, and a rainstorm provides the incoming photons (rain drops). Imagine a field covered with buckets aligned neatly in rows and columns throughout the entirety of the area (Figure 2.1). After the rainstorm (CCD integration), each bucket is transferred in turn and metered to determine the amount of water collected. A written record (final CCD image) of the amount of water in each bucket will thus provide a two-dimensional record of the rainfall within the field.

Referring to the actual mechanisms at work within a CCD, we start with the method of charge generation within a pixel: the photoelectric

Fig. 2.1. CCDs can be likened to an array of buckets that are placed in a field and collect water during a rainstorm. After the storm, each bucket is moved along conveyor belts until it reaches a metering station. The water collected in each field bucket is then emptied into the metering bucket within which it can be measured. From Janesick & Blouke (1987).

effect. Incoming photons strike the silicon within a pixel and are easily absorbed if they possess the correct wavelength (energy). Silicon has a band gap energy of 1.14 electron volts (eV), and so it easily absorbs light of energy 1.1 to 4 eV (11,000 to 3,000 Å).[†] Photon absorption causes the silicon to give up a valence electron and move it into the conduction band. Photons of energy 1.1 eV to near 4 or so eV generate single electron–hole pairs, whereas those of higher energy produce multiple pairs (see Section 2.2.7 and Chapter 7). Left to themselves, these conduction band electrons would recombine back into the valence level within approximately 100 microseconds. Silicon has a useful photoelectric effect range of 1.1 to about 10 eV, which covers the near-IR to soft X-ray region (Reicke, 1994). Above and below these limits, the CCD material appears transparent to the incoming photons.

Once electrons have been freed to the conduction band of the silicon, they must be collected and held in place until readout occurs. The details of the actual construction of each pixel within a CCD, that is, the formation of the MIS capacitor with its doped silicon, layers of silicon dioxide, etc., are beyond the scope of this book (Janesick & Elliott, 1992; Eccles, Sim, & Tritton, 1983), but suffice it to say that each pixel has a structure conducive to allowing applied voltages to be placed on sub-pixel sized electrodes called gates. These gate structures provide each pixel with the ability to collect the freed electrons and hold them in a potential well until the end of the exposure. In a typical arrangement, each pixel has associated with it three gates, each of which can be set to a different voltage potential. The voltages are controlled by clock circuits with every third gate connected to the same clock. Figure 2.2 illustrates this clocking scheme for a typical three-phase device.

We note in Figure 2.2 that, when an exposure ends, the clock voltages are manipulated such that the electrons that have been collected and held in each pixel's +10 volt potential well by clock voltage V3 can now be shifted within the device. Note that electrons created anywhere within the pixel during the exposure (where each pixel has a surface area equal to the total area under all three gates) will be forced to migrate toward the deepest potential well. When the exposure is terminated and CCD readout begins, the voltages applied to each gate are cycled (this process is called clocking the device) such that the charge stored within each pixel during the integration is electronically shifted.

[†] The energy of a photon of a given wavelength (in electron volts) is given by $E(eV) = 12{,}407/\lambda(\text{Å})$.

Fig. 2.2. Schematic voltage operation of a typical three-phase CCD. The clock voltages are shown at three times during the readout process, indicating their clock cycle of 0, 10, and 5 volts. One clock cycle causes the stored charge within a pixel to be transfered to its neighboring pixel. CCD readout continues until all the pixels have had their charge transfered completely out of the array and through the A/D converter. From Walker (1987).

A simple change in the voltage potentials (V3 goes to +5 volts, while V1 becomes +10 volts and so on) allows the charge to be shifted in a serial fashion along columns from one CCD pixel to another throughout the array. The transfer of the total charge from location to location within the array is not without losses. As we will see, each charge transfer (one of which occurs for each voltage change or clock cycle) has an associated efficiency. This efficiency value is the percent of charge transferred compared with that which was actually collected. Modern values for the charge transfer efficiency (CTE) are near 0.99999 (i.e., 99.999% efficient) for each transfer.

Each column in the array is connected in parallel and thus each pixel shift is mimiked throughout the entire array simultaneously. One clock cycle moves each row of pixels up one column, with the top row being shifted off the array into what is called the output shift register or horizontal shift register. This register is simply another row of pixels hidden from view (i.e., not exposed to incident light) and serves as the transition between active rows on the array and the output of the device. Once an entire row is shifted into the output register, and before any further row shifts on the active area occur, each pixel in the output register is shifted out one at a time (in a similar manner as before) into the output electronics. Here, the charge collected within each pixel is measured as a voltage and converted into an output digital number (see Section 2.4). Each pixel's collected charge is sensed and amplified by an output amplifier. CCD output amplifiers are designed to have low noise and are built directly into the silicon circuitry; thus they are often referred to as on-chip amplifiers. These amplifiers must work with extremely small voltages and are rated, as to their sensitivity, in volts per electron. Typical values are in the range of 0.5 to 4 microvolts per electron.

The output voltage from a given pixel is converted to a digital number (DN) and is typically discussed from then on as either counts or ADUs (analog-to-digital units). The amount of voltage needed (i.e., the number of collected electrons or received photons) to produce 1 ADU is termed the gain of the device. We will discuss the gain of a CCD in Chapter 3 and here only mention a few items of interest about it. A typical CCD gain might be 10 electrons/ADU, which means that for every 10 electrons collected within a pixel, the output from that pixel will produce, on average, a count or DN value of 1. For example, with this gain value if a pixel collects 1,000 electrons (photons), the output pixel value stored in the computer would be 100 ADUs. For 1,500 electrons 150 ADUs would be produced and for 17,234 electrons, the output pixel value would be 1,723 ADUs (note, not 1723.4). Digital output values can only be integer numbers and it is clear already that the discrimination between different pixel values can only be as good as the resolution of the gain and digital conversion of the device.

The process of conversion of the output voltage signal into a DN is performed within a device called an analog-to-digital converter (A/D or ADC). We will see later on that there is an intimate connection between the number of digital bits available in the A/D and the value of the gain that can or should be used for the CCD. The output DNs are usually stored initially in computer memory and then moved to tape or disk for later manipulation.

The process of shifting each entire CCD row into the output register, shifting each pixel along within this register, and finally performing the voltage conversion of each pixel's stored charge by the A/D to produce a DN value is continued until the entire array of pixels has been read out. For large-format, modern CCD arrays, this process may take upwards of a few minutes to complete a single readout of the entire device. Note that for a 2048 × 2048 CCD, the charge collected in the last pixel to be readout has to be transferred over four million times.

The array size of a single CCD, as well as the size of a given pixel on a device, is controlled by the current limitations of manufacturing. How large one can make a good quality, large-scale integrated circuit and how small one can make a MIS capacitor, both of which have demanding requirements for near perfect operation, set the scale of CCD and pixel sizes that are available. The interested reader is referred to Janesick & Elliott (1992) and Robinson (1988a) for further details.

2.2 CCD Types

When reading about CCDs, one of the most confusing issues can be the various terms listed in the literature or in commercial documents. Items such as backside illuminated, buried channel, and antiblooming are just a few. This section will provide a brief discussion to some of these terms while Chapter 3 will discuss CCD characteristics in detail. Further information can be found in the references listed in Appendix A. In particular, readers desiring a microscopic look at the electronic structures of a CCD integrated circuit are referred to the many SPIE articles listed. Some terms, such as quantum efficiency and full well capacity, will be used here without proper introduction. This will be rectified in the next chapter.

2.2.1 Surface Channel versus Buried Channel CCDs

As discussed above, charge stored within a pixel is moved from pixel to pixel during readout via changes in electrical potential between the biased gates. For surface channel CCDs, this charge movement occurs "on the surface" of the CCD, being transfered between overlapping gates. For example, each pixel in a three-phase CCD has three overlapping gates for each imaging cell or pixel. Each cell has an associated transfer efficiency and maximum possible transfer rate. The efficiency is near to, but not quite, 100% for each cell (we will discuss this charge-transfer efficiency (CTE) further in Chapter 3). Typical rates of transfer for CCD television cameras (or video cameras) are several megahertz, while low-light level applications with cooled devices (i.e., those occuring in astronomy) use rates nearer to the kilohertz range (Eccles, Sim, & Tritton, 1983).

The major drawback to the surface channel CCD is the presence of trapping states that occur at the boundaries of the gates. These traps, which are caused by imperfections within the silicon lattice or the gate structures, cause a loss of transfer efficiency by trapping some of the electrons from other pixels as they pass by during readout. Traps within CCDs are therefore undesirable for faint light level applications such as astronomical imaging. One method that can be used to eliminate traps (although not in general use anymore because of the advent of buried channel devices; see below) is to raise the entire surface charge level of the CCD above the level needed to fill in any traps. This is accomplished by illuminating the CCD at a low level prior to exposure of the astronomical source. This technique, called a pre-flash or a fat-zero, allows any nonlinearities resulting from traps at low light levels to be avoided while only slightly increasing the overall noise level of the resulting image (Djorgovski, 1984; Tyson & Seitzer, 1988).

Deferred charge is another source of nonlinearity sometimes present at low light levels (Gilliland, 1992; Baum, Thomsen, & Kreidl, 1981). Often referred to as charge skimming, charge depletion, or low light level nonlinearity, this particular worry has all but disappeared in modern, low readout noise devices. We will address nonlinearity in CCDs in a more general manner in Section 13.6.

A better solution than those just discussed is to move the charge from pixel to pixel during readout via a channel of semiconductor material that is away from the gate structures. This "buried channel" architecture results from the application of an additional layer of semiconductor material placed under the CCD surface. The buried channel greatly enhances the charge movement through the device by reducing the number of possible trap sites and by decreasing the transfer time between adjacent cells. Higher transfer rates (up to 100 MHz) and high CTE values (up to >99.995%) are easily accomplished with buried channel devices.

The price paid for the addition of the buried channel is that the total charge storage capacity for each pixel is reduced by 3 or 4 times that of a surface channel detector. However, since the operation at very low signal levels is much improved, the overall dynamic range and sensitivity of a buried channel device become much higher.

2.2.2 Front-Side and Back-Side Illuminated CCDs

CCDs are manufactured as single large-scale integrated devices. They have a front side, where all the gate structures and surface channel layers are deposited, and a back side, which is simply bulk silicon generally covered with a thin conductive layer of gold. CCDs used as front-side illuminated devices work as the name implies, that is, illumination occurs on the front of the CCD with the photons being absorbed by the silicon after passing directly through the surface gate structures. The device thickness is of order 300 microns from front to back making these chips relatively high in their susceptibility to detection of cosmic rays. Because the photons must first pass through the gate structures before they can be absorbed by the silicon, front-side illuminated CCDs have lower overall quantum efficiencies than the back-side devices (discussed below). However, front-side devices provide a flatter imaging surface and the actual CCD itself is easier to handle and work with. Figure 2.3 provides a schematic view of a single front-side illuminated pixel.

Back-side illuminated devices, also known as thinned devices, are again justly named. The CCD, after manufacture, is physically thinned to about 15 microns by various etching techniques (Lessor, 1994). The device is then mounted on a rigid substrate upside down and illuminated from behind. The incoming photons are now able to be absorbed directly into the bulk silicon pixels without the interference of the gate structures. The advantages in this type of CCD are that the relative quantum efficiency greatly exceeds that of a front-side device and the response of the detector to shorter wavelength light is improved since the photons

Fig. 2.3. Schematic view of a single front-side illuminated CCD pixel. The square labeled "front contact" is a representation of part of the overall gate structure. The letters "p" and "n" refer to regions within the pixel consisting of silicon doped with phosphorus and boron respectively.

no longer need to pass through the pixel gates. Disadvantages of back-side devices are in the areas of shallower pixel well depths (owing to the much smaller amount of material present per pixel), possible nonuniform thinning leading to surface and flat-field nonuniformities, and increased expense incurred by the thinning and mounting process.

In the next chapter, we will explore the idea of quantum efficiency further and provide an example of the large differences present in these two types of CCDs. Back-side devices generally have about twice the quantum efficiency for a given wavelength compared with front-side devices.

2.2.3 Interline and Frame Transfer CCDs

Interline transfer CCDs are specially constructed devices in which each column of imaging (active) pixels is paralleled by a light-shielded column of storage (or inactive) pixels. The device is used as a normal CCD but after the exposure ends, each light-sensitive column of data is quickly shifted into its neighboring light-shielded column. The shift occurs in only a few microseconds and so image smear is almost nonexistent and the need for a mechanical shutter is precluded. The shifted image data can then be clocked out of the device while the active columns are again integrating on the source of interest. Interline devices are used in many of the world's high-speed photon counting array detectors owing to their fast shift operation (Timothy, 1988).

The imaging area in an interline device is not continuous as each active column is paralleled by an inactive one; thus there is an immediate reduction in the overall extrinsic quantum efficiency by a factor of two. The electrode and gate structures of the active pixels are a bit different from those in a normal CCD, causing a further reduction in quantum efficiency and a somewhat different overall spectral response.

Frame transfer devices work in a manner similar to that of having two separate CCDs butted together. One half of the device is active (exposed to incoming light) and records the image while the other half is shielded and used as a storage device. The end of the integration sees the image data quickly transfered or shifted to the identical storage half, where it can be read out slowly as a new integration proceeds on the active side. Frame transfer CCDs are used in most commercial video and television cameras for which the active image is read out at video rates (30 frames per second). Astronomical imaging generally can not occur at such high rates because of photon starvation, but frame transfer devices have been used in a number of interesting instruments.

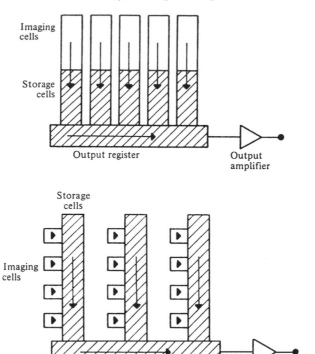

Fig. 2.4. Cartoon view of (top) a frame transfer CCD and (bottom) an interline CCD. From Eccles, Sim, & Tritton (1983).

Figure 2.4 provides an illustration of these two types of CCD. We note here that in both of these special purpose CCD types, the movement of the accumulated charge within each active area to the shielded, inactive area can be accomplished at very high rates. These rates are much higher than standard CCD readout rates because no sampling of the pixel charge or A/D conversion to an output number occurs until the final readout from the inactive areas. As we will see, the A/D process takes a finite time and is one cause of noise added to the output signal. Therefore, any on-chip shifting or summing of charge will introduce essentially no additional noise into the data.

2.2.4 Antiblooming CCDs

Observations of bright objects can lead to the collection of too many photoelectrons within one or more of the pixels of a CCD, causing the effected pixels to overflow.[†] This condition will occur when the full well capacity of a pixel (see next chapter) is exceeded. The result of such saturation is termed bleeding and is easily noted on a CCD image as one or more bright pixels with a bright streak leading away from the region, generally in two opposite directions.[‡] Saturation effects can be irritating or, even worse, the streak can bleed into an object of interest rendering its measurement useless. Figure 2.5 shows a CCD image with a saturated star.

To counter pixel saturation behavior, one could simply integrate for a shorter period of time. This will cause the offending object to supply less photons per integration and thus not saturate any CCD pixels. However, to obtain the same exposure level, a number of frames would have to be taken and co-added together later. We will see in the next chapter that this is not always equivalent (in noise terms) to a single long exposure. Therefore, although shorter exposures are a simple approach, they are not always practical. Many times we wish to study a field in which objects of various brightness are interspersed, requiring the need for considerable dynamic range within a single image. Thus, the antiblooming gate was invented (Neely & Janesick, 1993).

Antiblooming gates have added electrical structures manufactured into the CCD itself, allowing saturated pixels to be "drained" off and controlled instead of bleeding into neighboring pixels, destroying their integrity. The offending pixels, in such a case, are still not usable, but the neighboring pixels are. The trade-off that occurs for a CCD with antiblooming gates is the following: About 30% of the active pixel area is lost (as it becomes the drain gate) giving a remaining active pixel area of only 70%. The use of such a gate structure is therefore equivalent to an effective QE reduction. Such additional gate areas also reduce the

[†] Think here of the buckets in the field during a rainstorm and what would happen if too much water fell into any given one of them.
[‡] The reader should be aware of the fact that CCDs can saturate in two different ways. As discussed here, saturation can occur if a given pixel is exposed to too much light and overflows, exceeding its full well capacity. This type of saturation leads to the result discussed here (see Figure 2.5), that is, bleeding of excess charge into neighboring pixels. The other way in which a CCD can saturate is if any given pixel collects more charge than can be represented by the output from the A/D converter (see Sections 2.4 and 3.4), even if the amount collected is below the full well capacity of the pixel. This type of saturation does not lead to charge bleeding but yields such oddities as flat-topped star profiles or constant (exact) pixel values over the region of A/D saturation. As the amount of charge collected by the pixels within a CCD increases, eventually one of these two types of saturation will set the usable upper limit for a given system. We will see that the linearity of a CCD (Sections 3.4 and 3.6) may also limit the maximum usable ADU level of a given observation.

(a)   (b)

Fig. 2.5. Two equal-length CCD exposures of a bright star (SAO 110456). The normal CCD exposure (a) shows typical bleeding caused by saturation within the CCD. The CCD exposure on the right (b) was made with an antiblooming CCD and clearly shows the much reduced bleeding from the bright star. From Neely & Janesick (1993).

overall spatial resolution of any measurement (as the drain gates leave a gap between adjacent pixels) and lower the total pixel well depth by about one half. A rule of thumb here is that a typical antiblooming gate CCD requires about twice the integration time to reach the same signal level as a similar device without antiblooming gates.

Antiblooming gates are currently available on certain makes of CCD. When deciding if they are right for you, keep the following in mind: Longer exposures will be required for a given signal level, and if working in a spectral region of already low or reduced quantum efficiency, antiblooming gates will reduce the signal even further for a given integration time. On the plus side, if bleeding of bright objects will be a detriment to your science, the antiblooming gate option may be a good choice.

2.2.5 Multipinned Phase CCDs

A method developed to achieve ultra-low dark current in a CCD is that of multipinned phase operation or MPP mode. CCDs operating in MPP mode are common in inexpensive, "off-the-shelf" devices for which thermoelectric cooling techniques are exploited. The MPP design relies on the inversion of all three clock voltages in connection with additional semiconductor doping (usually boron or phosphorus) of the phase three gate. The effect of this semiconductor architecture is to greatly lower the dark (thermal) current and this can even allow room temperature operation if the light levels are sufficient. Generally, astronomical low light level applications require CCDs to be cooled to near $-100°$C before usable images can be obtained. Dark currents in MPP mode CCDs are near 0.01 electrons per second per pixel at temperatures of only -40 to $-65°$C, levels that are equal to non-MPP mode operation at $-100°$C.

MPP mode CCDs have allowed a new revolution in CCD technology. Items such as digital cameras (CCDs operating at room temperature) are now commonplace. The lower thermal noise, even at room temperature, allows integration and readout (for high light level scenes) to be used to produce digital photographs. While a boon to higher temperature operation, the full well depth of each pixel is reduced by two to three times that of the same device in non-MPP mode. The higher temperature operation must be balanced against the much reduced dynamic range of the modified CCD. New design strategies and higher quality silicon wafer technology are making progress in restoring the full well depth while keeping the advantages of MPP operation. A CCD modified to run in MPP mode has altered electronic design and pixel structures and, therefore, cannot be made to operate normally. Thus, an MPP CCD does not allow one to decide which mode of operation to use nor does it provide the ability to switch between MMP and normal operation.

2.2.6 Orthogonal Transfer CCDs

We have seen that a typical CCD has three gates per pixel and that the readout operation is via clocking the pixel charge in a three-phase fashion. The charge with a given pixel can only be shifted in the vertical direction, that is, from row to row, until it reaches the output register. A new type of CCD, the orthogonal transfer CCD (OTCCD), has been developed that allows each pixel's charge to be shifted both vertically and horizontally on the array (Burke et al., 1994).

The ability to move the charge in both directions within the OTCCD comes about through the use of a complex, four-phase mode of operation. Four gates are used in each pixel, two of which are triangular and make up the central part and two of which are rectangular but smaller in area and are split into pairs surrounding the pixel center. The larger number and size of the gates lead to a lower overall QE for an OTCCD and more detailed intra pixel QE effects are to be expected (Jorden, Deltorn, & Oates, 1994; Jorden, Deltorn, & Oates, 1993).

The first application of an OTCCD was to compensate for image motion during an integration, similar to a tip-tilt correction (Tonry, Burke, & Schechter, 1997). A 1024×512 OTCCD device was built to allow one half of the OTCCD to image, quickly read out, and centroid on a bright star, while the other half was used to integrate on a scene of interest. As the bright star center wandered during the integration, the object frame half of the OTCCD was electronically shifted (both vertically and horizontally) many thousand times per 100 seconds in order to move the image slightly (~0.5 arcsec) and follow the bright star centroid position. The final result from the OTCCD was an image with much improved seeing and little loss of information.

The complex gate structure causes small dead spots within each pixel as well as allowing charge to be trapped in numerous pockets within a pixel during charge shifting. These two effects amount to about 3% flux losses, which are probably compensated for in the final image. Advances in gate structure design and impurity control in the device composition have eliminated most of these losses (Tonry, Burke, & Schechter, 1997). While still early in development, the OTCCD promises to be extremely useful for certain applications in astronomical imaging.

2.2.7 Superconducting Tunnel Junctions

Although not truly a CCD, the superconducting tunnel junction (STJ) is a very recent semiconductor device addition to the field of astronomical imaging. Using a layer of polycrystalline niobium (Nb) supercooled to a temperature of only a few hundred milli kelvin, a STJ diode acts as a photon detector (Perryman et al., 1994). Used initially as an X-ray photon counter, the STJ is seeing new applications being developed in the areas of UV, visible, and IR light detection.

In Chapter 7, we will discuss how CCD detection of a high energy photon is accomplished and additionally how this detection and its associated photoelectron production provide information on the energy (wavelength) of the incoming light particle.[†] In general, light of low energy (UV to IR) does not produce multiple free electron pairs within a semiconductor; thus each incident photon (regardless of wavelength) looks like all the rest upon detection. Separation of spectral band must be accomplished by filtering the input beam.

Supercooled Nb STJ devices, however, allow UV, optical, and even IR photons to produce numerous pairs of free electrons, thus allowing a determination of the energy (wavelength) of each incoming particle. Recent application of a STJ device (Peacock et al., 1996) has shown sensitivity to wavelengths of 2,000 to 10,000 Å with a QE of near 50%. Spectral resolution of the incoming light is near 450 Å at present, but future devices made of other superconducting materials such as aluminum promise wavelength coverage from 10 Å to well over 1 micron at spectral resolutions of 80 Å.

The first astronomical test of a STJ occured during the fall of 1998 at the Observatorio del Roque de los Muchachos on La Palma in the Canary Islands. A single STJ diode, placed at the telescope focus, produced a broad band spectrum of a star by simply imaging it.

Superconducting tunnel junctions are a far way off from being a routine detector in use at an observatory. After all it took CCDs nearly ten years after their introduction to become commonplace. Thus, although not currently available for use, STJs are a remarkable device that will no doubt become an invaluable asset for the future of astronomy.

2.3 CCD Coatings

To make use of the properties of CCDs in wavebands for which silicon is not sensitive or to enhance performance at specific wavelengths, CCDs can and have been coated with various materials.[‡] These coating materials allow CCDs to become sensitive to photons normally too

---

[†] In the world of photon counting astronomy, using the number of photoelectrons produced to determine the wavelength (energy) of the incident photon is termed pulse-height spectroscopy.
[‡] Nonoptical operation of a CCD can also be accomplished in this manner. See Chapter 7.

blue to allow absorption by the silicon. They generally consist of organic phosphors that down-convert incident UV light into longer wavelength photons, easily detected by the CCD. These vacuum deposited coatings are often used with front-side illuminated, thick CCDs (although not always; the *Hubble Space Telescope* WF/PC I CCDs, which were thinned devices, were phosphor coated) and can be viewed as an inexpensive method of increasing blue response without the cost and complexity of thinning the device.

A coronene phosphor has been commonly used in recent years to convert photons shortward of 3,900 Å to photons at a wavelength near 5,200 Å. The use of such a coating causes a "notch" in the overall QE curve as the CCD QE is falling rapidly near 4,000 Å but coronene does not become responsive until 3,900 Å. Another common phosphor coating, Lumogen, eliminates this QE notch, as it is responsive to wavelengths between 500 and 4,200 Å (see Figure 2.6). An interesting side note is that Lumogen is the commercial phosphorescent dye used in yellow highlighting pens. The QE of a coated CCD may increase in the UV and shortward regions by amounts upwards of 15–20%. Transparent in the visible and near-IR regions, properly deposited coatings can also act as antireflection (AR) coatings for a CCD. More on the use of coated CCDs is contained in Chapter 7.

The main problem encountered when using coated CCDs is that the deposited material tends to evaporate off the CCD when under vacuum conditions (see Chapter 7). A less important problem is the loss of spatial resolution due to the finite emission cone of the light generated in the coating. The specifics of coating materials and the process used to coat CCDs are both beyond the scope of this book. Appendix A and Janesick & Elliott (1992), Lessor (1990), Schaeffer et al. (1990), and Schempp (1990) provide further readings in this area.

2.4 Analog-to-Digital Converters

Analog-to-digital (A/D) converters are not really a subject for this book. However, the output pixel values from a CCD (digital numbers stored in a computer) are completely determined by the method used to sample the charge on a pixel and how that sampling is converted into a digital number or data number to be stored in computer memory. A/D converters for use in CCDs, including linear and logarithmic types, have been reviewed in Opal (1988).

Fig. 2.6. The top plot shows QE curves for a *Hubble Space Telescope* WF/PC prototype CCD before and after being coated with Lumogen. Note the increased UV response of the coated CCD. The bottom plot shows the QE properties of a WF/PC prototype in the far-UV spectral region. Presented are two curves, one for a Coronene coated CCD and one for a Lumogen coated CCD. From Trauger (1990).

   The output from each pixel in a CCD must be examined by electronic means in order to determine how much charge has been collected during a given integration and to convert that charge value into a digital number. As each packet is clocked out of the CCD output register, it is passed through an on-chip amplifier circuit built directly into the CCD substrate. Next, the packet passes into an external low-noise solid state amplifier, to increase the signal strength, followed by input into the integrating amplifier. The output from the integrating amplifier is then sent directly into an A/D converter, which converts the analog voltage into a digital number to be stored in computer memory. The electronic circuitry just described, minus the on-chip amplifier, is often called the CCD head electronics and is usually located in a box very near or attached directly to the CCD dewar itself.

To measure the voltage of each charge packet, in preparation for conversion into a digital number, a difference measurement is performed between the charge packet itself and the sum of the charge packet plus a constant reset voltage. A well-regulated constant voltage source is needed in this step to supply the voltage to the electronics of the inverting plus integrating amplifier circuits. An integrating capacitor, connected across the integrating amplifier, first samples the reset voltage passed through the inverting amplifier for typically 20 μs. A single pixel charge packet is then passed through the noninverting amplifier with the output being sampled by the same integrating capacitor for the same time period. Equal time sampling is critical as one has to worry about charge decay in the capacitor. Generally, however, the RC time constant for the capacitor is of order a few seconds, while the sampling is a few tens of microseconds. Both signals sampled by the capacitor contain the reset voltage, but only one has the pixel charge packet. Thus the difference formed by the two opposite signed voltages provides a very good estimate of the actual charge level within the sampled pixel. The value of the reset voltage, whatever it may be, is completely eliminated by this differencing. This technique, for sampling the amount of charge within each pixel, is used in almost all current day CCDs and is called correlated double sampling (CDS). Details of the CDS process and some CDS variations such as the dual slope process and triple correlated sampling are discussed in Janesick & Elliott (1992), Opal (1988), Joyce (1992), and McLean (1997b).

The assignment of an output digital number to the value of the charge packet from each pixel is the job of the A/D converter. The charge stored in each pixel has an analog value (discrete only at the 1 electron level[†]) and the process of CDS decides how to assign each pixel's charge packet to a specific data number (DN) or analog-to-digital unit (ADU) (Merline

---

[†] Even though CCDs are called digital devices, the output from each pixel is really an analog voltage that must be measured and digitized by the A/D converter.

& Howell, 1995). As briefly mentioned in Chapter 1, this assignment is dependent on the device gain as follows.

If the gain for a CCD has a value of 10 electrons/ADU, the digitization process tries to divide each pixel's charge packet into units of 10 electrons each, assigning 1 ADU to each 10 electrons it measures. If, after division, there are less than 10 electrons left then the remaining charge is not counted and is lost and therefore unknown to the output pixel value. We will see (Sections 3.4 and Howell & Merline (1991)) that the gain of a device can have a great effect on its ability to provide good photometric information.

The ultimate readout speed of a given CCD depends on how fast the process of pixel examination and A/D conversion can take place. This entire process is called digitization. Modern large-format devices can, and often do, use two or four output amplifiers during readout (one at each device corner) providing a faster overall readout time. However, such readout schemes can introduce systematic offsets between each quadrant of the CCD owing to slight differences in the four output amplifiers and associated electronics. This effect is often seen in images produced by large-format CCDs, where one or more quadrants will show an apparently different background level. The readout speed of a CCD is also related to the number of bits of precision desired (A/D converters with fewer bits of precision work faster then those with more bits) and the acceptable readout noise level (faster pixel readout increases the introduced noise level). Current ultra-low noise A/D converters can distinguish the reference value from the collected charge in a given pixel at the 4 to 5 electron level.

Current CCD controllers (i.e., the hardware that controls the electronic functions of a CCD) are quite complex in design and yet simple in their use. Readout rates for CCDs are a prime concern as larger format chips may require excessively long readout times. Controllers such as those developed by San Diego State University (Leach, 1995) can read pixels out at a maximum rate of 10 μs/pixel, but a practical rate of near 50 μs/pixel is generally used to keep the read noise to a minimum (<10 electrons/pixel/read). Even at these seemingly fast rates (20 kHz), a 2048 × 2048 CCD, containing over four million pixels, takes over 200 seconds to read out.

3

Characterization of Charge-Coupled Devices

Even casual users of CCDs have run across the terms read noise, dark current, signal-to-noise, and many other possibly mysterious sounding bits of CCD jargon. This chapter will discuss the meanings of the terms used to characterize the properties of CCD detectors. Techniques and methods by which the reader can determine some of these properties on their own and why certain CCDs are better or worse for a particular application are discussed in the following chapters. Within the discussions, mention will be made of older types of CCDs. While these are generally not available or used anymore, there is a certain historical perspective to such a presentation and it will likely provide some amusement for the reader along the way.

One item to keep in mind throughout this chapter and in the rest of the book is that all electrons look alike. When a specific amount of charge is collected within a pixel during an integration, one can no longer know the exact source of each electron (i.e., was it due to a stellar photon or is it an electron generated by thermal motions within the CCD itself). We have to be clever to separate the signal from the noise. There are two notable quotes to keep in mind while reading this text. The first is from an early review article on CCDs by Craig Mackay (1986), who states: "The only uniform CCD is a dead CCD." The second is from numerous discussions I have had with CCD users and it is: "To understand your signal, you must first understand your noise."

3.1 Quantum Efficiency

The composition of a CCD is essentially pure silicon. This element is thus ultimately responsible for the response of the detector to various wavelengths of light. The wavelength dependence of silicon can be understood

26

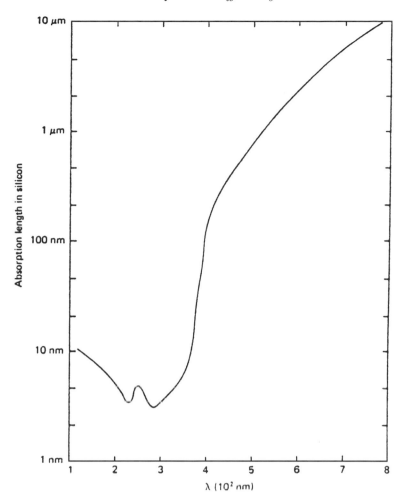

Fig. 3.1. The photon absorption length in silicon is shown as a function of wavelength in nanometers. From Reicke (1994).

in an instant by glancing at Figure 3.1. Shown here is the length of silicon needed for a photon of a specific wavelength to be absorbed. Absorption length is defined as the distance for which 63% $(1/e)$ of the incoming photons will be absorbed. Figure 3.1 clearly shows that, for light outside the range of about 3,500 to over 8,000 Å, the photons (1.) pass right through the silicon, (2.) get absorbed within the thin surface layers or gate structures, or (3.) simply reflect off the CCD surface. At short wavelengths, 70% or more of the photons are reflected, and for very short wavelengths (as for long wavelengths) the CCD becomes completely transparent. Thus the quantum effeciency of a typical CCD device will approximately mirror the photon absorption curve for silicon. Shortward of ~2,500 Å (for thinned devices) or about 25 Å (for thick devices) the detection probability for photons increases again. However, due to their much higher energy, these photons lead to the production of multiple electron–hole pairs within the silicon and may also produce damage to the CCD itself (see Chapter 7).

CCD quantum efficiencies are therefore very dependent on the thickness of the silicon that intercepts the incoming photons. This relation between absorption probability and CCD thickness is why front-side illuminated devices are more red sensitive (the photons have a higher chance of absorption) and why they have lower overall (blue) QEs (since the gate structures can be close to or even exceed the necessary absorption depths of as small as only a few atomic layers). A few front-side CCDs have been produced with special gate structures that are transparent to incoming blue and UV photons. In thinned devices, the longer wavelength photons are likely to pass right through the CCD without being absorbed at all.

Figure 3.2 provides some examples of typical quantum efficiency curves. Note the large difference in QE between thinned and thick CCDs. Quantum efficiency or QE curves allow one to quickly evaluate the relative collecting power of the device as a function of wavelength. QE curves shown in the literature are generally assumed to be representative of each and every pixel on the device, that is, all pixels of a given device are assumed to work identically and have the same wavelength response. This is almost true, but it is the "almost" that makes flat fielding of a CCD necessary. In addition, the QE curves shown or delivered for a particular device may be only representative of a "typical" device of the same kind, but they may not be for the exact device of interest.

Placing an antireflection (AR) coating on a CCD (both for visible and near-UV light) increases the QE and extends the range of good QE well into the 3,000 Å region. The need for AR coatings comes from the fact that silicon, like most metallic substances, is a good reflector of visible light. If you ever have a chance to hold a CCD, you will easily see just how well the surface does indeed reflect visible light. All the QE curves in Figure 3.2 have the overall shape expected based on the absorption properties of silicon as shown in Figure 3.1. Graphical illustrations of QE curves almost always include photon losses due to the gate structures, electron recombination within the bulk silicon itself, surface reflection,

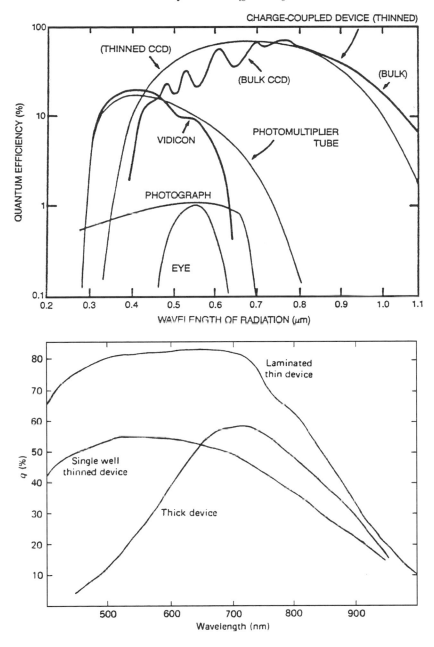

Fig. 3.2. The top panel shows QE curves for various devices indicating why CCDs are a quantum leap above all previous imaging devices. The failure of CCDs at wavelengths shorter than about 3,500 Å has been completely eliminated via thinning or coating of the devices. The bottom plot show representative QE curves for both thick and thinned devices. Laminated and single-well devices are two processes used to produce mechanically stable thinned CCDs.

and, for very long or short wavelengths, losses due to the almost complete lack of absorption by the CCD. Given that all these losses are folded together into the QE value for each wavelength, it should be obvious that changes within the CCD structure itself (such as radiation damage or operating temperature changes) can cause noticeable changes in its quantum efficiency.

Measurement of the quantum efficiency of a CCD is usually performed with the aid of complicated laboratory equipment including well-calibrated photodiodes. Light at each wavelength is used to illuminate both the CCD and the photodiode, and the relative difference in the two readings is recorded. The final result of such an experiment is an absolute QE curve for the CCD (with respect to the calibrated diode) over the range of all measured wavelengths.

To measure a CCD QE curve yourself, a few possibilities exist. You may have access to a setup such as that described above. Measurements can also be made at the telescope itself. One good method of QE measurement for a CCD consists of employing a set of narrow-band filters and a few spectrophotometric standard stars. Performing such a task will provide a good relative QE curve and, if one knows the filter and telescope throughput well, a good absolute QE curve. A detailed reference as to what is involved in the measurement of a spectrophotometric standard star is provided by Tüg et al. (1977). When producing a QE curve by using the above idea, the narrow-band filters provide wavelength selection while the standard stars provide a calibrated light source. A less ambitious QE curve can be produced using typical broad-band (i.e., Johnson) filters, but the final result is not as good because of the large bandpasses and throughput overlap of some of the filters. In between a detailed laboratory setup and the somewhat sparse technique of using filters at the telescope, another method exists. Using an optics bench, a calibrated light source covering the wavelength range of interest, and some good laboratory skills, one can produce a very good QE curve for a CCD and can even turn the exercise into a challenging classroom project.

3.2 Readout Noise

Readout noise, or just read noise, is usually quoted for a CCD in terms of the number of electrons introduced per pixel into your final signal upon readout of the device. Read noise consists of two inseparable components. First is the conversion from an analog signal to a digital number, which is not perfectly repeatable. Each on-chip amplifier and A/D circuit will produce a statistical distribution of possible answers centered on a mean value.[†] Thus, even for the hypothetical case of reading out the same pixel twice, each time with identical charge, a slightly different answer may be produced. Second, the electronics themselves will introduce spurious electrons into the entire process, yielding unwanted random fluctuations in the output. These two effects combine to produce an additive uncertainty in the final output value for each pixel. The average (one sigma) level of this uncertainty is the read noise and is limited by the electronic properties of the on-chip output amplifier and the output electronics (Djorgovski, 1984).[‡]

The physical size of the on-chip amplifier, the integrated circuit construction, the temperature of the amplifier, and the sensitivity (generally near 1–4 μV/detected photon, i.e., collected photoelectron) all contribute to the read noise for a CCD. In this micro world, the values for electronic noise are highly related to the thermal properties of the amplifier, which in turn determines the sensitivity to each small output voltage. Amazing as it seems, the readout speed, and thus the rate at which currents flow through the on-chip amplifier, can cause thermal swings in the amplifier temperature, which can affect the resulting read noise level. Generally, slower readout speeds produce lower read noise but this reduced readout speed must be weighed against the overall camera duty cycle. Small effects caused by amplifier heating can even occur between the readout of the beginning and end of a single CCD row, as the last charge packets pass through a slightly hotter circuit. Increasing the physical size of the already small microamplifiers can alleviate these small temperature swings, but larger amplifiers have a higher input capacitance, thereby lowering the sensitivity of the amplifier to small voltages.

Additional work on amplifier design, methods of clocking out pixels, and various semiconductor doping schemes can be used to improve the performance of the output electronics. CCD manufacturers invest large efforts into balancing these issues to produce very low read noise devices. Many details of the various aspects of read noise in CCDs are discussed by Janesick & Elliott (1992).

In the output CCD image, read noise is added into every pixel every time the array is read out. This means that a CCD with a read noise of

---

[†] The distribution of these values is not necessarily Gaussian (Merline & Howell, 1995).
[‡] We note here that the level of the read noise measured, or in fact any noise source within a CCD, can never be smaller than the level of discretization produced by the A/D converter (see Sections 2.4 and 3.6).

20 electrons will, on average, contain 20 extra electrons of charge in each pixel upon readout. High read noise CCDs are thus not very good to use if co-addition of two or more images is necessary. The final resultant image will not be quite as good as one long integration of the same total time, as each co-added image will add in one times the read noise to every pixel in the sum. However, for modern CCDs (see Section 4.4), read noise values are very low and are hardly ever the dominant noise with which one must be concerned. Good read noise values in today's CCDs are in the range of 10 electrons per pixel per read or less. These values are far below read noise levels of ten years ago, which were as high as 50–100 electrons, and even those are well down from values of 500 or more electrons/pixel/read present in the first astronomical CCDs.

In Section 4.3, we will discuss a simple method by which one may determine for oneself the read noise of a given CCD. This determination can be performed with any working CCD system and does not require special equipment, removal of the CCD from the camera dewar, or even removal from the telescope.

3.3 Dark Current

Every material at a temperature much above absolute zero will be subject to thermal noise within. For silicon in a CCD, this means that when the thermal agitation is high enough, electrons will be freed from the valence band and become collected within the potential well of a pixel. When the device is read out, these dark current electrons become part of the signal, indistinguishable from astronomical photons. Thermal generation of electrons in silicon is a strong function of the temperature of the CCD, which is why astronomical use generally demands some form of cooling (McLean, 1997b). Figure 3.3 shows a typical CCD dark current curve, which relates the amount of thermal dark current to the CCD operating temperature. Within the figure the theoretical relation for the rate of thermal electron production is given.

Dark current for a CCD is usually specified as the number of thermal electrons generated per second per pixel or as the actual current generated per area of the device (i.e., picoamps cm^{-2}). At room temperature, the dark current of a typical CCD is near 2.5×10^4 electrons/pixel/second. Typical values for properly cooled devices range from 2 electrons per second per pixel down to very low levels of approximately 0.04 electrons per second for each pixel. Although 2 electrons of thermal noise generated within a pixel every second sounds very low, a

Fig. 3.3. Experimental (symbols) and theoretical (line) results for the dark current generated in a typical three-phase CCD. The rate of dark current, in electrons generated within each pixel every second, is shown as a function of the CCD operating temperature. E_g is the band gap energy for silicon. From Robinson (1988a).

typical 15 minute exposure of a faint astronomical source would include 1,800 additional (thermal) electrons within each CCD pixel upon readout. These additional charges cannot, of course, be uniquely separated from the photons of interest after readout. The dark current produced in a CCD provides an inherent limitation on the noise floor of a CCD. Because dark noise has a Poisson distribution the noise actually introduced by thermal electrons into the signal is proportional to the square root of the dark current (see Section 4.4).

Cooling of CCDs is generally accomplished by one of two methods. The first, and usually the one used for scientific CCDs at major observatories, is via the use of liquid nitrogen (or in some cases liquid air). The CCD and associated electronics (the ones on or very near the actual CCD itself, called the head electronics) are encased in a metal dewar under vacuum. Figure 3.4 shows a typical astronomical CCD dewar

Fig. 3.4. A typical CCD dewar. This is the Mark-II Universal dewar originally produced in 1984 at Kitt Peak National Observatory. The dewar held 1.75 liters of liquid nitrogen providing a CCD operating time of approximately 12 hours between fillings. This dewar could be used in up-looking, down-looking, and side-looking orientations. From Brar (1984).

(Brar, 1984; Florentin-Nielsen, Anderson, & Nielsen, 1995). The liquid nitrogen (LN2) is placed in the dewar and, although not in direct physical contact with the CCD, cools the device to temperatures of near $-100°C$. Since LN2 itself is much colder than this, CCDs are generally kept at a constant temperature ($\pm0.1°C$) with an on-board heater.

In fact, the consistency of the CCD temperature is very important as the dark current is a strong function of temperature (Figure 3.3) and will vary considerably due to even modest changes in the CCD temperature.

A less expensive and much less complicated cooling technique makes use of thermoelectric cooling methods. These methods are employed in essentially all "off-the-shelf" CCD systems and allow operation at temperatures of -20 to $-50°$C or so, simply by plugging the cooler into an electrical outlet. Peltier coolers are the best known form of thermoelectric cooling devices and are discussed in Martinez & Klotz (1998). CCD operation and scientific quality imaging at temperatures near $-30°$C is possible, even at low light levels, owing to advances in CCD design and manufacturing techniques and the use of multipinned phase operation (see Chapter 2). Other methods of cooling CCDs that do not involve LN2 are discussed in McLean (1997a).

The amount of dark current a CCD produces depends primarily on its operating temperature, but there is a secondary dependence upon the bulk properties of the silicon used in the manufacture. Even CCDs produced on the same silicon wafer can have slightly different dark current properties. As with most of the noise properties of a given CCD, custom tailoring the CCD electronics (such as the bias level and the readout rate) can produce much better or much worse overall dark current and noise performance.

3.4 CCD Pixel Size, Pixel Binning, Full Well Capacity, and Windowing

This section is a combination of a few related topics concerning the amount of charge that can be stored within a given pixel during an integration. We have seen that CCD thinning, MPP operation, and small physical pixel size all place limitations on the total number of electrons that can be collected within a pixel. The general rule of thumb is that the physically larger the pixel (both in area and in thickness) the more charge that it can collect and store.

The amount of charge a pixel can hold in routine operation is termed its full well capacity. A Kodak CCD with 9-micron pixels (meaning 9 microns on a side for the projected area, but giving no indication of the thickness of the CCD) operating in MPP mode has a full well capacity per pixel of 85,000 electrons. In contrast, a SITe CCD with 24-micron pixels can have a full well capacity per pixel of over 350,000 electrons.

When we discussed the method by which a CCD is read out (Chapter 2) it was stated that each row is shifted in turn into the output register and then digitized, and the resulting DN value is sent off to the computer. During this process, each pixel's value is increased on average by one times the read noise. If we could add up the charge within say 4 pixels before they are digitized, we would get a final signal level equal to $\sim$4 times each single pixel's value, but only one times the read noise. This process is called on-chip binning and, if selected, occurs prior to readout within the CCD output register (Merline & Howell, 1995; Smith, 1990b). Pixels can be binned (summed) in both vertical and horizontal directions. "On-chip" means that the accumulated charge from each pixel involved in the binning is brought together and summed before the process of A/D conversion occurs. This summing process is done in the output register and is limited by the size of the "pixels" within this register. Generally, the output register pixels can hold five to ten times the charge of a single active pixel. This deeper full well capacity of the output register pixels allows pixel summing to take place.

Older CCD systems that allowed on-chip binning had plug boards mounted on the sides of the dewar. Certain combinations of the plug wires produced different on-chip binning patterns and users could change these to suit their needs. Today, most CCDs have the ability to perform pixel summing as a software option (Leach, 1995). Binning terminology states that normal operation (or "high resolution" as it is called by many low-cost CCDs) is a readout of the CCD in which each pixel is read, digitized, and stored. This is called 1×1 binning or unbinned operation. Binning of 2×2 would mean that an area of 4 adjacent pixels will be binned or summed on-chip within the output register during readout, but before A/D conversion. The result of this binning operation will produce only one "superpixel" value, which is digitized and stored in the final image; the original values in each of the four summed pixels are lost forever. Mainly for spectroscopic operation, binning of 3×1 is commonly used, with the 3 being in the direction perpendicular to the dispersion. Binning of CCD pixels decreases the image resolution, usually increases the final signal-to-noise value of a measurement, and reduces the total readout time and final image size. For example, a 1024×1024 CCD binned 2×2 will have a final image size of only 512×512 pixels and the readout time will be reduced by about a factor of four.

Pixel binning gives flexibility to the user for such applications as (using a high binning factor) quick readout for focus tests, nights with poor seeing, or very low surface brightness observations. Spectroscopic observations with a CCD, high spatial resolution imaging, or bright object observations will benefit from the use of a low binning factor. Binning factors that are very large (say 40 × 40 pixels) might be of use in some rare cases, but they will be limited by the total amount of charge one can accumulate in a single superpixel of the output register.

A related function available with some CCDs is "windowing." Windowing allows the user to choose a specific rectangular region (or many regions) within the active area of the CCD to be read out upon completion of the integration. The CCD window is often specified by providing the operating software with a starting row and column number and the total number of x, y pixels to use. For example, using a 2048 × 2048 CCD to make high speed imaging observations would be difficult, but windowing the CCD to use only the first 512 rows and columns (0, 0, 512, 512) allows for much faster readout and requires far less storage for the image data. Of course, your object of interest must now be positioned within these first 512 rows and columns, and not at the center of the CCD as may be usual. Other applications of CCD windowing would include choosing a cosmetically good subregion of a large CCD or only a rectangular strip to read out from a larger square CCD, when making spectroscopic observations. CCD windowing is independent of any on-chip binning, and one can both window and bin a CCD for even more specific observational needs.

3.5 Overscan and Bias

In an attempt to provide an estimate of the value produced by an empty or unexposed pixel within a CCD, calibration measurements of the bias level can be used.[†] Bias or zero images allow one to measure the zero noise level of a CCD. For an unexposed pixel, the value for zero collected photoelectrons will translate, upon readout and A/D conversion, into a mean value with a small distribution about zero.[‡] To avoid negative

[†] For more on bias frames and their use in the process of CCD image calibration, see Chapter 4.
[‡] Before bias frames, and in fact before any CCD frame is taken, a CCD should undergo a process known as "wiping the array." This process makes a fast read of the detector, without A/D conversion or data storage, in order to remove any residual dark current or photoelectron collection that may have occurred during idle times between obtaining frames of interest.

numbers in the output image,[†] CCD electronics are set up to provide a positive offset value for each accumulated image. This offset value, the mean "zero" level, is called the bias level. A typical bias level might be a value of 400 ADU (per pixel), which, for a gain of 10 e^-/ADU, equals 4,000 electrons. This value might seem like a large amount to use, but historically temporal drifts in CCD electronics due to age, temperature, or poor stability in the electronics, as well as much higher read noise values, necessitated such levels.

To evaluate the bias or zero noise level and its associated uncertainty, specific calibration processes are used. The two most common ones are: (1.) overscan regions produced with every object frame or (2.) usage of bias frames. Bias frames amount to taking observations without exposure to light (shutter closed), for a total integration time of 0.000 seconds. This type of image is simply a readout of the unexposed CCD pixels through the on-chip electronics, through the A/D converter, and then out to the computer producing a two-dimensional bias or zero image.

Overscan strips, as they are called, are a number of rows or columns (usually 32) or both that are added to and stored with each image frame. These overscan regions are not physical rows or columns on the CCD device itself but additional pseudo-pixels generated by sending additional clock cycles to the CCD output electronics. Both bias frames and overscan regions are techniques that allow one to measure the bias offset level and, more importantly, the uncertainty of this level.

Use of overscan regions to provide a calibration of the zero level generally consists of determining the mean value within the overscan pixels and then subtracting this single number from each pixel within the CCD object image. This process removes the bias level pedestal or zero level from the object image and produces a bias-corrected image. Bias frames provide more information than overscan regions, as they represent any

---

[†] Representation of negative numbers requires a sign bit to be used. This bit, number 15 in a 16-bit number, is 0 or 1 depending on whether the numeric value is positive or negative. For CCD data, sacrificing this bit for the sign of the number leaves one less bit for data, thus reducing the overall dynamic range. Therefore, most good CCD systems do not make use of a sign bit. One can see the effects of having a sign bit by viewing CCD image data of high numeric value but displayed as a signed integer image. For example, a bright star will be represented as various grey levels, but at the very center (i.e., the brightest pixels) the pixel values may exceed a number that can be represented by 14 bits (plus a sign). Once bit 15 is needed, the signed integer representation will be taken by the display as a negative value and the offending pixels will be displayed as black. This is due to the fact that the very brightest pixel values have made use of the highest bit (the sign bit) and the computer now believes the number is negative and assigns it a black (negative) greyscale value. This type of condition is discussed further in Appendix C.

two-dimensional structure that may exist in the CCD bias level. Two-dimensional (2-D) patterns are not uncommon for the bias structure of a CCD, but these are usually of low level and stable with time. Upon examination of a bias frame, the user may decide that the 2-D structure is nonexistent or of very low importance and may therefore elect to perform a simple subtraction of the mean bias level value from every object frame pixel. Another possibility is to remove the complete 2-D bias pattern from the object frame using a pixel-by-pixel subtraction (i.e., subtract the bias image from each object image). When using bias frames for calibration, it is usually best to work with an average or median frame composed of many (10 or more) individual bias images (Gilliland, 1992). This averaging eliminates cosmic rays,[†] read noise variations, and random fluctuations which will be a part of any single bias frame.

Variations in the mean zero level of a CCD are known to occur over time and are usually slow drifts over many months or longer, not noticeable changes from night to night or image to image. These latter types of changes indicate severe problems with the readout electronics and require correction before the CCD image data can be properly used.

Producing a histogram of a typical averaged bias frame will reveal a Gaussian distribution with the mean level of this distribution being the bias level offset for the CCD. We show an example of such a bias frame histogram in Figure 3.5. The width of the distribution shown in Figure 3.5 is related to the read noise of the CCD (caused by shot noise variations in the CCD electronics (Mortara & Fowler, 1981)) and the device gain by the following expression:

$$\sigma_{\text{ADU}} = \frac{\text{Read Noise}}{\text{Gain}}.$$

3.6 CCD Gain and Dynamic Range

The gain of a CCD is set by the output electronics and determines how the amount of charge collected in each pixel will be assigned to a digital number in the output image. Gain values are usually given in terms of the number of electrons needed to produce one ADU step within the A/D converter. Listed as electrons/Analog-to-Digital Unit (e^{-}/ADU), common gain values range from 1 (photon counting) to 150 or more.

[†] Cosmic rays are not always cosmic! They can be caused by weakly radioactive materials used in the construction of CCD dewars (Florentin-Nielsen, Anderson, & Nielsen, 1995).

Fig. 3.5. Histogram of a typical bias frame showing the number of pixels vs. each pixel ADU value. The mean bias level offset or pedestal level in this Loral CCD is near 1,017 ADU, and the distribution is very Gaussian in nature with a FWHM value of near 2 ADU. This CCD has a read noise of 10 electrons and a gain of 4.7 e$^-$/ADU.

One of the major advantages of a CCD is that it is linear in its response over a large range of data values. Linearity means that there is a simple linear relation between the input value (charge collected within a each pixel) and the output value (digital number stored in the output image).

The largest output number that a CCD can produce is set by the number of bits in the A/D converter. For example, if you have a 14-bit A/D, numbers in the range from zero to 16,383 can be represented.[†] A 16-bit A/D would be able to handle numbers as large as 65,535 ADU.

Figure 3.6 provides a typical example of a linearity curve for a CCD. In this example, we have assumed a 15-bit A/D converter capable of producing output DN values in the range of 0 to 32,767 ADU, a device gain of 4.5 e$^-$/ADU, and a pixel full well capacity of 150,000 electrons. The linearity curve shown in Figure 3.6 is typical for a CCD, revealing

---

[†] The total range of values that a specific number of bits can represent equals $2^{(number\ of\ bits)}$, e.g., $2^{14} = 16,384$. CCD output values are zero based, that is, they range from 0 to $2^{(number\ of\ bits)} - 1$.

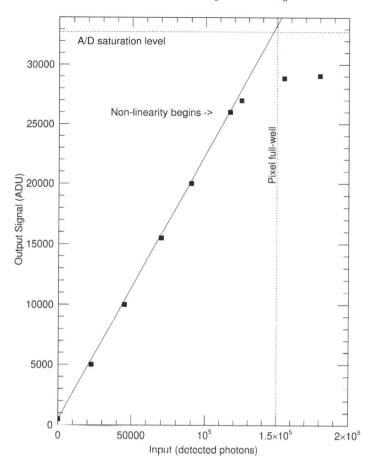

Fig. 3.6. CCD linearity curve for a typical three-phase CCD. We see that the device is linear over the output range from 500 ADU (the offset bias level of the CCD) to 26,000 ADU. The pixel full well capacity is 150,000 electrons and the A/D converter satuartion is at 32,767 ADU. In this example, the CCD nonlinearity is the limiting factor of the largest usable output ADU value. The slope of the linearity curve is equal to the gain of the device.

that over most of the range the CCD is indeed linear in its response to incoming photons. Note that the CCD response has the typical small bias offset (i.e., the output value being nonzero even when zero incident photons occur), and the CCD becomes nonlinear at high input values. For this particular CCD, nonlinearity sets in near an input level of 1.17×10^5 photons (26,000 ADU), a number still well within the range of possible output values from the A/D.

The user must be aware of three possible factors that can limit the largest usable output pixel value in a CCD image: the two types of saturation that can occur (A/D saturation and exceeding a pixel's full well capacity; see Sections 2.2.4 and 2.4) and nonlinearity. For the CCD in the example shown in Figure 3.6, A/D saturation would occur at an output value of $32{,}767 \cdot 4.5 = 147{,}451$ input photons. The pixel full well capacity is 150,000 electrons; thus pixel saturation will occur at a value of 33,333 ADU (150,000/4.5). Both full well and A/D saturations would produce noticeable effects in the output data such as bleeding or flat-topped stars. This particular example, however, illustrates the most dangerous type of situation that can occur in a CCD image. The nonlinear region, which starts at 26,000 ADU, is entered into before either type of saturation can occur. Thus, the user could have a number of nonlinear pixels (for example the peaks of bright stars) and be completely unaware of it. No warning bells will go off and no flags will be set in the output image to alert the user to this problem. The output image will be happy to contain (and the display will be happy to show) these nonlinear pixel values and the user, if unaware, may try to use such values in the scientific analysis.

Thus, it is very important to know the linear range of your CCD and to be aware of the fact that some pixel values, even though they are not saturated, may indeed be within the nonlinear range and therefore unusable. Luckily, most professional grade CCDs reach one of the two types of saturation before they enter their nonlinear regime. Be aware, however, that this is almost never the case with low quality, inexpensive CCD systems that tend to use A/Ds with fewer bits, poor quality electronics, or low grade (impure) silicon. Most observatories have linearity curves available for each of their CCDs and some manufacturers include them with your purchase.[†] If uncertain of the linear range of a CCD, it is best to measure it yourself.

One method of obtaining a linearity curve for a CCD is to observe a field of stars covering a range of brightness. Obtain exposures of say 1, 2, 4, 8, 16, etc. seconds, starting with the shortest exposure needed to provide good signal-to-noise ratios (see Section 4.4) for most of the stars and ending when one or more of the stars begins to saturate. Since you have obtained a sequence that doubles the exposure time for each frame, you should also double the number of incident photons collected per star in each observation as well. Plots of the output ADU values

[†] A caution here is that the supplied linearity curve may only be representative of your CCD.

for each star versus the exposure time will provide you with a linearity curve for your CCD.

A common, although not always good, method of determining the value to use for the CCD gain, is to relate the full well capacity of the pixels within the array to the largest number that can be represented by your CCD A/D converter. As an example, we will use typical values for a Loral 512 × 1024 CCD in current operation at the Royal Greenwich Observatory. This CCD has 15-micron pixels and is operated as a backside illuminated device with a full well capacity of 90,000 electrons per pixel. Using a 16-bit A/D converter (output values from 0 to 65,535) we could chose the gain as follows. Take the total number of electrons a pixel can hold and divide it by the total ADU values that can be represented: 90,000/65,536 = 1.37. Therefore, a gain choice of 1.4 e$^-$/ADU would allow the entire dynamic range of the detector to be represented by the entire range of output ADU values. This example results in a very reasonable gain setting, thereby allowing the CCD to produce images that will provide good quality output results.

As an example of where this type of strategy would need to be carefully thought out, consider a CCD system designed for a spacecraft mission in which the A/D converter only had 8 bits. A TI CCD was to be used, which had a full well capacity of 100,000 electrons per pixel. To allow imagery to make use of the total dynamic range available to the CCD, a gain value of 350 ($\sim$100,000/2^8) e$^-$/ADU was used. This gain value certainly made use of the entire dynamic range of the CCD, allowing images of scenes with both shadow and bright light to be recorded without saturation. However, as we noted before, each gain step is discrete, thereby making each output ADU value uncertain by $\pm$ the number of electrons within each A/D step. A gain of 350 e$^-$/ADU means that each output pixel value has an associated uncertainty of $\pm$1 ADU, which is equal to, in this case, $\pm$350 electrons, a large error if precise measurements of the incident flux are desired. The uncertainty in the final output value of a pixel, which is caused by the discrete steps in the A/D output, is called digitization noise and is discussed in Merline & Howell (1995).

To understand digitization noise let us take, as an example, a CCD that can be operated at two different gain settings. If we imagine the two gain values to be either 5 or 200 e$^-$/ADU and that a particular pixel collects 26,703 electrons (photons) from a source, we will obtain output values of 5,340 and 133 ADU respectively. Remember, A/D converters output only integer values and so any remainder is lost. In this example,

3 and 103 electrons respectively are lost as the result of the digitization noise of the A/D. More worrisome than this small loss of incident light is the fact that while each ADU step in the gain equals 5 e⁻/ADU case can only be incorrect by <5 electrons, the gain equals 200 e⁻/ADU case will be uncertain by ±200 electrons in each output ADU value. Two hundred electrons/pixel may not seem like much but think about trying to obtain a precise flux measurement for a galaxy that covers thousands of pixels on a CCD image or even a star that may cover tens of pixels. With an error of ±200 electrons/pixel multiplied by tens or many more pixels, the value of a galaxy's surface brightness at some location or similarly a stellar magnitude would both be highly uncertain.

The gain of a particular CCD system is set by the electronics and is generally not changeable by the user or there may be but a few choices available as software options. Often, major observatories allow the user to pre-select, during the proposal process, which CCD (and thereby which gain, well depth, etc.) to use during the observing run. Table 3.1 gives an example of the choices one has if applying for time at Kitt Peak National Observatory to perform spectroscopy. Note that the A/D converter in all cases is 16 bits and that one of the CCDs reaches nonlinearity before it reaches A/D saturation. Thus, some care must be taken in deciding which CCD and gain combination you should use to accomplish your science objectives.

How A/D converters actually determine the assignment of the number to output for each pixel and whether the error in this choice is equally distributed within each ADU step is a detailed matter of interest but lies outside the scope of this book. A discussion of how the digitization noise affects the final output results from a CCD measurement is given in Merline & Howell (1995) and a detailed study of ADCs used for CCDs is given in Opal (1988).

In our above discussion of the gain of a CCD, we mentioned the term dynamic range a few times but did not offer a definition. The dynamic range of any device is the total range over which it operates or for which it is sensitive. For audio speakers this number is usually quoted in decibels, and this tradition is used for CCDs as well. The dynamic range of a CCD is given by the expression

$$D(\text{dB}) = 20 \times \log_{10}(\text{full well capacity/readnoise}).$$

A CCD with a full well capacity of 100,000 electrons per pixel and a read noise of 10 electrons would have a dynamic range of 80 dB.

Table 3.1. *Spectrographs and CCDs available at Kitt Peak*

Instrument	CCD	Scale ("/pixel)	Pixel Size (μ)	Gain (e−/ADU)	Read Noise (e−)	A/D limit (e−)	Linearity (e−s at 0.1%)
4-m RC Spec	Tek 2K × 2K	0.59	24	3.1	4.0	201,500	220,000
4-m CryoCam	Loral 1K × 1K	0.34	15	0.8	15	52,000	150,000
2.1-m GoldCam	Loral 2K × 2K	0.78	15	1.4	8.5	91,000	80,000
WIYN Hydra	Tek 2K × 2K	0.78	24	1.7	4.3	110,500	190,000

Table 3.2. *Various typical CCD properties*

	RCA	Fairchild	TI	EEV	Thomson	Kodak	Lorel (Ford)	Tektronix (SITe)
Pixel Format	320 × 512	380 × 488	800 × 800	2048 × 4096	1024 × 1024	765 × 510	3072 × 1024	1024 × 1024
Pixel Size (μ)	30	18 × 30	15	13.5	19	9	15	24
Detector Size (mm)	10 × 15	7 × 15	12 × 12	28 × 55	20 × 20	7 × 4.5	46 × 15	25 × 25
Full Well Capacity (e−)	350,000	>200,000	50,000	180,000	500,000	85,000	>140,000	>170,000
Dark Current (e−/pixel/hr)	40	>1000	16	<2	22	1800	11	<5
Illumination	Front	Front	Back	Back	Front	Front	Back	Back
Peak QE (%)	70	12	70	85	40	40	90	75
Read Noise (e−)	80	>150	15	7	5	12	9	6
CTE	0.99995	0.99975	0.999985	0.99999	0.99996	0.99997	0.99997	0.999999
Operating Temp. (°C)	−100	−100	−120	−120	−110	−35	−120	−111
Gain (e−/ADU)	13.5	50	5	1.1	5	2.3	1.2	1.3
Readout Time (s)	45	65	70	185	45	45	142	300

3.7 Summary

This chapter has concentrated on defining the terminology commonly used when discussing CCDs. The brief nature of this book does not allow the many more subtle effects, such as deferred charge, cosmic rays, or pixel traps, to be discussed further nor does it permit any discussion of the finer points of each of the above items. The reader seeking a deeper understanding of the details of CCD terminology (a.k.a., someone with a lot of time on his or her hands) is referred to the references given in this chapter and the detailed reading list in Appendix A. Above all, the reader is encouraged to find some CCD images and a workstation capable of image processing and image manipulation and to spend a few hours of time exploring the details of CCDs for themselves.

As a closing thought for this chapter, Table 3.2 provides a sample of the main properties of various CCDs. (Also see Appendix B.) The table lists CCDs that were used in the early 1980s up to the best available chips today. The sample shown tries to present the reader with an indication of the typical properties exhibited by CCDs. Included are those of different dimension, of different pixel size, having front and back illumination, cooled by LN2 or thermoelectrically, and those available from different manufacturers. A scan of Table 3.2 allows one to see at a glance many of the corresponding relations we have just discussed, for example, the pixel well depth variation with front (thick) or back (thin) illuminated devices. In 1980, three companies made commercially available CCDs: RCA, GEC, and Fairchild. By 1994, over thirty commercially produced CCDs were on the market as well as at least five CCDs produced only for defense related work. Table 3.2 will serve us later on in the book when we are in need of typical CCD values for examples or calculations.

4

CCD Imaging

This chapter will deal with the most basic use of a CCD, that of direct imaging. We will discuss a few more preliminaries such as flat fields, the calculation of gain and read noise for a CCD, and how the signal-to-noise value for a measurement is determined. The chapter then continues by providing a brief primer on the use of calibration frames in standard two-dimensional CCD data image reduction. Finally, we cover some aspects of CCD imaging itself including applications of wide-field imaging with CCD mosaics and CCD drift scanning.

4.1 Image or Plate Scale

One of the basic parameters of importance to a CCD user is that of knowing the plate scale of your image. Plate scale is a term that originated when photographic plates were used as the main imaging device and is often given in arcsec/mm. For a CCD user, however, a more convenient unit for the plate scale is arcsec/pixel. Clearly the conversion from one to the other is simple.

The focal ratio of a telescope is given by

$$f/ = \frac{\text{focal length of primary mirror}}{\text{primary mirror diameter}},$$

where both values are in the same units and "primary mirror" would be replaced by "primary objective lens" for a refractor. Taking the focal length of the primary (f) in mm and the CCD pixel size (μ) in microns, we can calculate the CCD plate scale as

$$P = \frac{206{,}265 \times \mu}{1{,}000 \times f} \quad (\text{arcsec/pixel}),$$

where 206,265 is the number of arcseconds in 1 radian and 1,000 is the conversion factor between millimeters and microns.

For a 1-m telescope of $f/ = 7.5$, the focal length (f) of the primary would be 7,500 mm. If we were to use a Loral CCD with 15-micron pixels as an imager, the above expression would yield an image scale on the CCD of 0.41 arcsec/pixel. This image scale is usually quite a good value for direct imaging applications for which the seeing is near 1 or so arcseconds.

There are times, however, when the above expression for the plate scale of a CCD may not provide an accurate value. This could occur if there are additional optics within the instrument that change the final f-ratio in some unknown manner. Under these conditions, or simply as an exercise to check the above calculation, one can determine the CCD plate scale observationally. Using a few CCD images of close optical double stars with known separations, measurement of the center positions of the two stars and application of a bit of plane geometry will allow an accurate determination of the pixel-to-pixel spacings, and hence the CCD plate scale.

4.2 Flat Fielding

To CCD experts, the term "flat field" can cause shivers to run up and down their spine. For the novice, it is just another term to add to the lexicon of CCD jargon. If you are in the latter category, don't be put off by these statements but you might want to take a minute and enjoy your thought of "How can a flat field be such a big deal?" In principle, obtaining flat field images and flat fielding a CCD image are conceptually easy to understand, but in practice the reality that CCDs are not perfect imaging devices sets in.

The idea of a flat field image is simple. Within the CCD, each pixel has a slightly different gain or QE value when compared with its neighbors. In order to flatten the relative response for each pixel to the incoming radiation, a flat field image is obtained and used to perform this calibration. Ideally, a flat field image would consist of uniform illumination of every pixel by a light source of identical spectral response to that of your object frames. That is, you want the flat field image to be spectrally and spatially flat. Sounds easy, doesn't it? Once a flat field image is obtained, one then simply divides each object frame by it and voilá: instant removal of pixel-to-pixel variations.

Before talking about the details of the flat fielding process and why it is not so easy, let us look at the various methods devised to obtain flat field exposures with a CCD. All of these methods involve a light source that is brighter than any astronomical image one would observe. This light source provides a CCD calibration image of high signal-to-noise ratio. For imaging applications, one very common procedure used to obtain a flat field image is to illuminate the inside of the telescope dome (or a screen mounted on the inside of the dome) with a light source, point the telescope at the bright spot on the dome, and take a number of relatively short exposures so as to not saturate the CCD. Since the pixels within the array have different responses to different colors of light, flat field images need to be obtained through each filter that is to be used for your object observations. As with bias frames discussed in the last chapter, five to ten or more flats exposed in each filter should be obtained and averaged together to form a final or master flat field, which can then be used for calibration of the CCD. Other methods of obtaining a CCD flat field image include taking CCD exposures of the dawn or dusk sky or obtaining spatially offset images of the dark night sky; these can then be median filtered to remove any stars that may be present (Gilliland, 1992; Tobin, 1993; Massey & Jacoby, 1992; Tyson, 1990).

To allow the best possible flat field images to be obtained, many observatories have mounted a flat field screen on the inside of each dome and painted this screen with special paints (Massey & Jacoby, 1992) that help to reflect all incident wavelengths of light as uniformly as possible. In addition, most instrument user manuals distributed by observatories discuss the various methods of obtaining flat field exposures that seem to work best for their CCD systems. Illumination of dome flat field screens has been done by many methods, from a normal 35 mm slide projector, to special "hot filament" quartz lamps, to various combinations of lamps of different color temperature and intensity mounted like headlamps on the front of the telescope itself. Flat fields obtained by observation of an illuminated dome or dome screen are referred to as dome flats, while observations of the twilight or night sky are called sky flats.

CCD imaging and photometric applications use dome or sky flats as a means of calibrating out pixel-to-pixel variations. For spectroscopic applications, flat fields are obtained via illumination of the spectrograph slit with a quartz or other high intensity projector lamp housed in an integrating sphere (Wagner, 1992). The output light from the sphere attempts to illuminate the slit, and thus the grating of the spectrograph, in a similar manner to that of the astronomical object of interest. This type of flat field image is called a projector flat. While the main role of a flat field image is to remove pixel-to-pixel variations within the CCD, these calibration images will also compensate for any image vignetting and for time-varying dust accumulation, which may occur on the dewar window and/or filters within the optical path.

Well, so far so good. So what is the big deal about flat field exposures? The problems associated with flat field images and why they are a topic discussed in hushed tones in back rooms may still not be obvious to the reader. There are two major concerns. One is that uniform illumination of every CCD pixel (spatially flat) to one part in a thousand is often needed but in practice is very hard to achieve. Second, QE variations within the CCD pixels are wavelength dependent. This wavelength dependence means that your flat field image should have the exact wavelength distribution over the bandpass of interest (spectrally flat) as that of each and every object frame you wish to calibrate. Quartz lamps and twilight skies are not very similar at all in color temperature (i.e., spectral shape) to that of a dark nighttime sky filled with stars and galaxies.[†] Sky flats obtained of the dark nighttime sky would seem to be our savior here, but these types of flat fields require long exposures to get the needed signal-to-noise ratio and multiple exposures with spatial offsets to allow digital filtering (e.g., median) to be applied in order to remove the stars. In addition, the time needed to obtain (nighttime) sky flats is likely not available to the observer who generally receives only a limited stay at the telescope. Thus, whereas very good calibration data lead to very good final results, the fact is that current policies of telescope scheduling often mean that we must somehow compromise the time used for calibration images with that used to collect the astronomical data of interest.

Within the above detailed constraints on a flat field image, it is probably the case that obtaining a perfect, color-corrected flat field is an impossibility. But all is not lost. Many observational projects do not require total perfection of a flat field over all wavelengths or over the entire two-dimensional array. Stellar photometry resulting in differential measurements or on-band/off-band photometry searching for particular emission lines are examples for which one only needs to have good flat field information over small spatial scales on the CCD. However, a project with end results of absolute photometric calibration over large spatial extents (e.g., mapping of the flux distribution within the spiral arms of an extended galaxy) does indeed place stringent limits on flat fielding requirements. For such demanding observational programs, some observers have found that near-perfect flats can be obtained through the

[†] One good sky region for twilight flats has been determined to be an area 13 degrees east of zenith just after sunset (Chromey & Hasselbacher, 1996).

use of a combination of dome and sky flats. This procedure combines the better color match and low-spatial frequency information from the dark night sky with the higher signal-to-noise, high spatial frequency information of a dome flat. Experimentation to find the best method of flat fielding for a particular telescope, CCD, and filter combination, as well as for the scientific goals of a specific project, is highly recommended.

A summary of the current best wisdom on flat fields depends on who you talk to and what you are trying to accomplish with your observations. The following advice is one person's view.

What does the term "a good flat field" mean? An answer to that question is: A good flat field allows a measurement to be transformed from its instrumental values into numeric results in a standard system that results in an answer that agrees with other measurements made by other observers. For example, if two observers image the same star, they both observe with a CCD using a V filter, and they each end up with the final result of $V = 14.325$ magnitudes in the Johnson system then, assuming this is an accurate result, one may take this as an indication of the fact that each observer used correct data reduction and analysis procedures (including their flat fielding) for the observations.

The above is one way to answer the question, but it still relies on the the fact that observers need to obtain good flat fields. Without them, near perfect agreement of final results is unlikely. While the ideal flat field would uniformly illuminate the CCD such that every pixel would receive equal amounts of light in each color of interest, this perfect image is generally not produced with dome screens, the twilight sky, or projector lamps within spectrographs. This is because good flat field images are all about color terms. That is, the twilight sky is not the same color as the nighttime sky, neither of which are the same color as a dome flat. If you are observing red objects, you need to worry more about matching the red color in your flats; for blue objects you worry about the blue nature of your flats. Issues to consider include the fact that if the Moon is present, the sky is bluer then when the Moon is absent, dome flats are generally reddish owing to their illumination by a quartz lamp of relatively low filament temperature, and so on. Thus, just as in photometric color transformations, the color terms in flat fields are all important. One needs to have a flat field that is good, as described above, plus one that also matches the colors of interest to the observations at hand.

Proper techniques for using flat fields as calibration images will be discussed in Section 4.5. Appendix A offers further reading on this subject and the material presented in Djorgovski (1984), Sterken (1995),

Tyson (1990), and Gudehus (1990) is of particular interest concerning flat fielding techniques.

4.3 Calculation of Read Noise and Gain

We have talked about bias frames and flat field images in the text above and now wish to discuss the way in which these two types of calibration data may be used to determine the read noise and gain for a CCD.

Noted above, when we discussed bias frames, was the fact that a histogram of such an image (see Figure 3.5) should produce a Gaussian distribution with a width related to the read noise and the gain of the detector. Furthermore, a similar relation exists for the histogram of a typical flat field image (see Figure 4.1). The mean level in the flat field shown in Figure 4.1 is $\bar{F} = 6{,}950$ ADU and its width (assuming it is

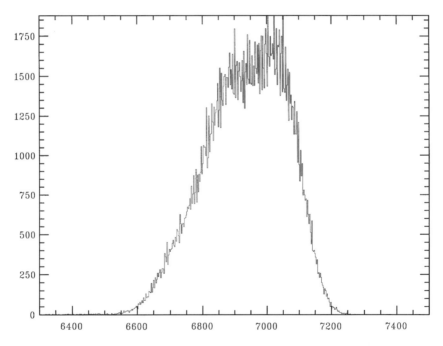

Fig. 4.1. Histogram of a typical flat field image. Note the fairly Gaussian shape of the histrogram and the slight tail extending to lower values. For this R-band image, the filter and dewar window were extremely dusty leading to numerous out of focus "doughnuts" (see Figure 4.4), each producing lower than average data values.

perfectly Gaussian (Massey & Jacoby, 1992)) will be given by

$$\sigma_{ADU} = \frac{\sqrt{F} \cdot Gain}{Gain}.$$

We have made the assumption in this formulation that the Poisson noise of the flat field photons themselves is much greater than the read noise. This is not unreasonable at all given the low values of read noise in present day CCDs.

Let us now look at how bias frames and flat field images can be used to determine the important CCD properties of read noise and gain. Using two bias frames and two flat field images, designated 1 and 2, we can proceed as follows. Determine the mean pixel value within each image.[†] We will call the mean values of the two bias frames $\bar{B}_1$ and $\bar{B}_2$ and likewise $\bar{F}_1$ and $\bar{F}_2$ will be the corresponding values for the two flats. Next, create two difference images ($B_1 - B_2$ and $F_1 - F_2$) and measure the standard deviation of these image differences: $\sigma_{B_1-B_2}$ and $\sigma_{F_1-F_2}$. Having done that, the gain of your CCD can be determined from the following:

$$Gain = \frac{(\bar{F}_1 + \bar{F}_2) - (\bar{B}_1 + \bar{B}_2)}{\sigma^2_{F_1-F_2} - \sigma^2_{B_1-B_2}},$$

and the read noise can be obtained from

$$Read\ Noise = \frac{Gain \cdot \sigma_{B_1-B_2}}{\sqrt{2}}.$$

4.4 Signal-to-Noise Ratio

Finally we come to one of the most important sections in this book, the calculation of the signal-to-noise (S/N) ratio for observations made with a CCD.

Almost every article written that contains data obtained with a CCD and essentially every observatory user manual about CCDs contain some version of an equation used for calculation of the S/N of a measurement. S/N values quoted in research papers, for example, do indeed give the reader a feel for the level of goodness of the observation (i.e., a S/N of

---

[†] Be careful here not to use edge rows or columns, which might have very large or small values owing to CCD readout properties such as amplifier turn on/off (which can cause spikes). Also, do not include overscan regions in the determination of the mean values.

100 is probably good while a S/N of 3 is not), but rarely do the authors discuss how they performed such a calculation.

The equation for the S/N of a measurement made with a CCD is given by

$$\frac{S}{N} = \frac{N_*}{\sqrt{N_* + n_{\mathrm{pix}}(N_S + N_D + N_R^2)}},$$

unofficially named the "CCD Equation" (Mortara & Fowler, 1981). Various formulations of this equation have been produced (e.g., Newberry (1991) and Gullixson (1992)), all of which yield the same answers of course, if used properly. The "signal" term in the above equation, N_*, is the total number of photons[†] (signal) collected from the object of interest. N_* may be from one pixel (if determining the S/N of a single pixel as sometimes is done for a background measurement), or N_* may be from several pixels, such as all of those contained within a stellar profile (if determining the S/N for the measurement of a star) or N_* may even be from say a rectangular area of X by Y pixels (if determining the S/N in a portion of the continuum of a spectrum).

The "noise" terms in the above equation are the square roots of N_*, plus n_{pix} (the number of pixels under consideration for the S/N calculation) times the contributions from N_S (the total number of photons per pixel from the background or sky), N_D (the total number of dark current electrons per pixel), and N_R^2 (the total number of electrons per pixel resulting from the read noise.[‡]

For those interested in more details of each of these noise terms, how they are derived, and why each appears in the CCD Equation, see Merline & Howell (1995). In our short treatise, we will remark on some of the highlights of that paper and present an improved version of the CCD Equation. However, let's first make sense out of the equation just presented.

For sources of noise that behave under the auspices of Poisson statistics (which includes photon noise from the source itself), we know that for a signal level of N, the associated 1 sigma error (1σ) is given by

---

[†] Throughout this book, we have and will continue to use the terms photons and electrons interchangeably when considering the charge collected by a CCD. In optical observations, every photon that is collected within a pixel produces a photoelectron; thus they are indeed equivalent. When talking about observations, it seems logical to talk about star or sky photons, but for dark current or read noise discussions, the number of electrons measured seems more useful.

[‡] Note that this noise source is not a Poisson noise source but a shot noise; therefore it enters into the noise calculation as the value itself, not the square root of the value as Poisson noise sources do.

$\sqrt{N}$. The above equation for the S/N of a given CCD measurement of a source can thus be seen to be simply the signal (N_*) divided by the summation of a number of Poisson noise terms. The n_{pix} term is used to apply each noise term on a per pixel basis to all of the pixels involved in the S/N measurement and the N_R term is squared since this noise source behaves as shot noise, rather than being Poisson-like (Mortara & Fowler, 1981). We can also see from the above equation that if the total noise for a given measurement $\sqrt{N_* + n_{\mathrm{pix}}(N_S + N_D + N_R^2)}$ is dominated by the first noise term, N_* (i.e., the noise contribution from the source itself), then the CCD Equation becomes

$$\frac{S}{N} = \frac{N_*}{\sqrt{N_*}} = \sqrt{N_*},$$

yielding the expected result for a measurement of a single Poisson behaved value.

This last result is useful as a method of defining what is meant by a "bright" source and a "faint" source. As a working definition, we will use the term bright source to mean a case for which the S/N errors are dominated by the source itself (i.e., $S/N \approx \sqrt{N_*}$), and we will take a faint source to be the case in which the other error terms are of equal or greater significance compared with N_*, and therefore the complete error equation (i.e., the CCD Equation) is needed.

The CCD equation above provides the formulation for a S/N calculation given typical conditions and a well-behaved CCD. For some CCD observations, particularly those that have high background levels, faint sources of interest, poor spatial sampling, or large gain values, a more complete version of the error analysis is required. We can write the complete CCD Equation (Merline & Howell, 1995) as

$$\frac{S}{N} = \frac{N_*}{\sqrt{N_* + n_{\mathrm{pix}}\left(1 + \frac{n_{\mathrm{pix}}}{n_B}\right)\left(N_S + N_D + N_R^2 + G^2\sigma_f^2\right)}}.$$

This form of the S/N equation is essentially the same as that given above, but two additional terms have been added. The first term, $(1 + n_{\mathrm{pix}}/n_B)$, provides a measure of the noise incurred as a result of any error introduced in the estimation of the background level on the CCD image. The term n_B is the total number of background pixels used to estimate the mean background (sky) level. One can see that small values of n_B will introduce the largest error as they will provide a poor estimate of the mean level of the background distribution. Thus, very large values of n_B are to be preferred but clearly some trade-off must be made between providing a good estimate of the mean background level and the use of pixels from areas on the CCD image that are far from the source of interest or possibly of a different character.

The second new term added into the complete S/N equation accounts for the error introduced by the digitization noise within the A/D converter. From our discussion of the digitization noise in Chapter 3, we noted that the error introduced by this process can be considerable if the CCD gain has a large value. In this term, $G^2\sigma_f^2$, G is the gain of the CCD (in electrons/ADU) and σ_f^2 is an estimate of the 1 sigma error introduced within the A/D converter[†] and has a value of approximately 0.289 (Merline & Howell, 1995).

In practice for most CCD systems in use and for most observational projects, the two additional terms in the complete S/N equation are often very small error contributors and can be ignored. In the instances for which they become important – for example, cases in which the CCD gain has a high value (e.g., 100 electrons/ADU), the background level can only be estimated with a few pixels (e.g., less than 200), or the CCD data are of poor pixel sampling (see Section 5.6) – ignoring these additional error terms will lead to an overestimation of the S/N value obtained from the CCD data.

Let us work through an example of a S/N calculation given the following conditions. A 300 second observation is made of an astronomical source with a CCD detector attached to a 1-m telescope. The CCD is a Thomson 1024 × 1024 device with 19-micron pixels and it happens that in this example the telescope has a fast f-ratio such that the plate scale is 2.6 arcsec/pixel.[‡] From Table 3.2 we find that, for this particular CCD, the read noise is 5 electrons/pixel/read, the dark current is 22 electrons/pixel/hour, and the gain (G) is 5 electrons/ADU. Using 200 background pixels surrounding our object of interest from which to estimate the mean background sky level, we take a mean value for N_B of 620 ADU/pixel. We will further assume here (for simplicity) that the CCD image scale is such that our source of interest falls completely within 1 pixel (good seeing!) and that after background subtraction (see Section 5.1), we find a value for N_* of 24,013 ADU. Ignoring the two additional minor error terms discussed above (as the gain is very small

---

[†] The parameter σ_f^2 and its value depend on the actual internal electrical workings of a given A/D converter. We assume here that for a charge level that is half way in between two output ADU steps (that is, 1/2 of a gain step), there is an equal chance that it will be assigned to the lower or to the higher ADU value when converted to a digital number. See Merline & Howell (1995) for further details.

[‡] Using the results from Section 4.1, what would the f-ratio of this telescope be?

and $n_B = 200$ is quite sufficient in this case), we can write the CCD Equation as

$$\frac{S}{N} = \frac{24{,}013(\text{ADU}) \cdot G}{\sqrt{24{,}013(\text{ADU}) \cdot G + (1) \cdot (620(\text{ADU}) \cdot G + 1.8 + 5^2(e^-))}}.$$

Note that all of the values used in the calculation of the S/N are in electrons, *not* in ADUs. The S/N value calculated for this example is ~342, a very high S/N. With such a good S/N measurement, one might suspect that this is a bright source. If we compare $\sqrt{N_*}$ with all the remaining error terms, we see that indeed this measurement has its noise properties dominated by the Poisson noise from the source itself and the expression $S/N \sim \sqrt{N_*} = 346$ works well here.

While the S/N of a measurement is a useful number to know, at times we would prefer to quote a standard error for the measurement as well. Using the fact that $S/N = 1/\sigma$, where σ is the standard deviation of the measurement, we can write

$$\sigma_{\text{magnitudes}} = \frac{1.0857\sqrt{N_* + p}}{N_*}.$$

In this expression, p is equal to $n_{\text{pix}}(1 + n_{\text{pix}}/n_B)(N_S + N_D + N_R^2 + G^2\sigma_f^2)$, the same assumptions apply concerning the two "extra" error terms, and the value of 1.0857 is the correction term between an error in flux (electrons) and that same error in magnitudes (Howell, 1993). We again see that if the Poisson error of N_* itself dominates, the term p can be ignored and this equation reduces to that expected for a 1σ error estimate in the limiting case of a bright object.

Additionally, one may be interested in a prediction of the S/N value likely to be obtained for a given CCD system and integration time. N_* is really $N \cdot t$, where N is the count rate in electrons (photons) per second (for the source of interest) and t is the CCD integration time. Noting that the integration time is implicit in the other quantities as well, we can write the following (Massey, 1990):

$$\frac{S}{N} = \frac{Nt}{\sqrt{Nt + n_{\text{pix}}(N_S t + N_D t + N_R^2)}},$$

in which we have again ignored the two minor error terms. This equation illustrates a valuable rule of thumb concerning the S/N of an observation: $S/N \propto \sqrt{t}$, not to t itself. Solving the above expression for t we find

$$t = \frac{-B + (B^2 - 4AC)^{1/2}}{2A},$$

where $A = N^2$, $B = -(S/N)^2 \, (N + n_{\text{pix}}(N_S + N_D))$, and $C = -(S/N)^2$ $n_{\text{pix}}N_R^2$. Most instrument guides available at major observatories provide tables that list the count rate expected for an ideal star (usually 10th magnitude and of 0 color index) within each filter and CCD combination in use at each telescope. Similar tables provide the same type of information for the observatory spectrographs as well. The tabulated numeric values, based on actual CCD observations, allow the user, via magnitude, seeing, or filter width, to scale the numbers to a specific observation and predict the S/N expected as a function of integration time.

4.5 Basic CCD Data Reduction

The process of standard CCD image reduction makes use of a basic set of images that form the core of the calibration and reduction process (Gullixson, 1992). The types of images used are essentially the same (although possibly generated by different means) in imaging, photometric, and spectroscopic applications. This basic set of images consists of three calibration frames – bias, dark, and flat field – and the data frames of the object(s) of interest. Table 4.1 provides a brief description of each image type and Figures 4.2–4.5 show examples of typical bias, dark, flat field, and object CCD images. Note that a CCD dark frame contains not only information on the level and extent of the dark current but also includes bias level information as well.

The use of the basic set of calibration images in the reduction of CCD object frames is as follows. First, subtract a mean bias frame (or dark frame if needed[†]) from your object frame. Then, divide the resulting image by a (bias subtracted) mean flat field image. That's all there is to it! These two simple steps have corrected your object frame for bias level, dark current (if needed), and nonuniformity within each image pixel. During the analysis of your object frames, it is likely that the background or sky contribution to the image will need to be removed or accounted for in some manner. This correction for the background sky

---

[†] The need for dark frames instead of simply bias frames depends entirely on the level of dark current expected during an integration or the stability of the dark current from integration to integration. The first situation depends on the operating temperature of the CCD. LN2 systems have essentially zero dark current, and thus bias frames are all that is needed. Inexpensive and thermoelectrically cooled CCD systems fall into the category of generally always needing dark frames as part of the calibration process.

Table 4.1. *Types of CCD images*

CCD Image Type	Image Description
Bias	This type of CCD image has an exposure time of zero seconds. The shutter remains closed and the CCD is simply readout. The purpose of a bias or zero frame is to allow the user to determine the underlying noise level within each data frame. The bias value in a CCD image is usually a low spatial frequency variation throughout the array, caused by the CCD on-chip amplifiers. This variation should remain constant with time. The rms value of the bias level is the CCD read noise. A bias frame contains both the DC offset level (overscan) and the variations on that level. The nature of the bias variations for a given CCD are usually column-wise variations, but may also have small row-wise components as well. Thus, a 2-D, pixel-by-pixel subtraction is often required. A single bias frame will not sample these variations well in a statistical fashion, so an average bias image of 10 or more single bias frames is recommended.
	★ ★ ★ ★ ★
Dark	CCD dark frames are images taken with the shutter closed but for some time period, usually equal to that of your object frames. That is, if one is planning to dark correct a 45 second exposure, a 45 second dark frame would be obtained. Longer dark frames can often be avoided using the assumption that the dark current increases linearly with time and a simple scaling can be applied. However, this is not always true. Dark frames are a method by which the thermal noise (dark current) in a CCD can be measured. They also can give you information about bad or "hot" pixels that exist as well as provide an estimate of the rate of cosmic ray strikes at your observing site. Observatory class CCD cameras are usually cooled with LN2 to temperatures at which the dark current is essentially zero. Many of these system therefore do not require the use of a dark exposure CCD frames in the calibration process. Thermoelectrically cooled systems are not cooled to low enough temperatures such that one may ignore the dark current. In addition, these less expensive models often have poor temperature stability allowing the dark current to wander a bit with time. Multiple darks averaged together are the best way to produce the final dark calibration frame. Note that if dark frames are used, the bias level of the CCD is present in them as well, and therefore separate bias frames are not needed.

(*Cont.*)

Table 4.1. *(Continued)*

CCD Image Type	Image Description
Flat Field	Flat field exposures are used to correct for pixel-to-pixel variations in the CCD response as well as any nonuniform illumination of the detector itself. Flat fields expose the CCD to light from either a dome screen, the twilight sky, the nighttime sky, or a projector lamp in an attempt to provide a high S/N, uniformly illuminated calibration image. For narrow-band imaging, flats are very helpful in removing fringing which may occur in object frames. Flat field calibration frames are needed for each color, wavelength region, or different instrumental setup used in which object frames are to be taken. A good flat should remain constant to about 1%, with 2% or larger changes being indicators of a possible problem. As with the other calibration frames, at least 5 or more flat fields should be taken and averaged to produce the final flat used for image calibration.
	★ ★ ★ ★ ★
Object	These are the frames containing the astronomical objects of interest. They are of some exposure length from 1 second or less up to many hours, varying for reasons of type of science, brightness of object, desired temporal sampling, etc. Within each object image pixel is contained contributions from the object and/or sky, read noise, thermally generated electrons, and possibly contributions from cosmic rays. Each pixel responds similarly but not exactly to the incident light, so nonuniformities must be removed. All of the noise and spatial factors are correctable to very low levels via standard CCD reductions as described in the text.

level in your image frames is performed as part of each specific analysis step using "sky" regions in the object frame itself and is not removed or corrected for with some sort of separate "sky" frame. In equational form, the calibration process can be written as

$$\text{Final Reduced Object Frame} = \frac{\text{Raw Object Frame} - \text{Bias Frame}}{\text{Flat Field Frame}},$$

where, again, the flat field image has already been bias subtracted and the bias frame would be replaced by a dark frame when appropriate.

Fig. 4.2. Shown is a typical CCD bias frame. The histrogram of this image was shown in Figure 3.5. Note the overall uniform structure of the bias frame.

4.6 CCD Imaging

This section details issues related to the application of using CCDs to produce images of an extended area of the sky. Examples of this type of CCD observation are multicolor photometry of star clusters, galaxy imaging to isolate star forming regions within spiral arm structures, deep wide-field searches for quasars, and extended low surface brightness mapping of diffuse nebulae. Use of the areal nature of a CCD introduces some additional issues related to the calibration procedures and the overall cosmetic appearance as any spurious spatial effects will have implications on the output result. We briefly discuss here a few new items that are of moderate concern in two-dimensional imaging and then move on to the topic of wide-field imaging with CCD mosaics.

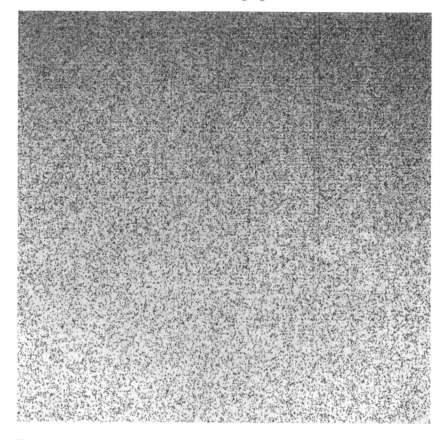

Fig. 4.3. Shown is a typical CCD dark frame. This figure shows a dark frame for a Kodak CCD operating in MPP mode and thermoelectrically cooled. Notice the nonuniform dark level across the CCD, being darker (greater ADU values) on the top. Also notice the two prominent partial columns with higher dark counts, which extend from the top toward the middle of the CCD frame. These are likely to be column defects in the CCD that occured during manufacture, but with proper dark subtraction they are of little consequence. The continuation of the figure shows the histogram of the dark frame. Most of the dark current in this 180 second exposure is uniformly distributed near a mean value of 180 ADU with a secondary maximum near 350 ADU. The secondary maximum represents a small number of CCD pixels that have nearly twice the dark current of the rest, again most likely due to defects in the silicon lattice. As long as these increased dark current pixels remain constant, they are easily removed during image calibration.

Fig. 4.3. (*Continued*)

4.6.1 CCD Fringing and Cosmetic Effects

We mentioned earlier that observations of monochromatic (or nearly so) light can cause a pattern of fringes to occur on a CCD image. These fringes, which are essentially Newton's rings, are caused by interference between light waves that reflect within the CCD or long wavelength light that passes through the array and reflects back into the array. Fringing may occur for CCD observations in the red part of the optical spectrum, when narrow-band filters are used, or if observations are made of a spectral regime (e.g., the I-band) that contains strong narrow emission lines.

The troubling aspect with fringing in CCD data is that it is often the case that the fringe pattern does not occur in the flat field frames or the pattern is highly variable throughout the night. Without a pattern match between the flats and the image data, fringe removal will not occur during image calibration, and residual fringes will remain in the final object images. One of the major causes of CCD fringing is the night sky emission lines that occur in the Earth's upper atmosphere (Pecker, 1970). These night sky lines are mainly attributed to OH transitions in the atmosphere and are powered by sunlight during the day. Since they are forbidden transitions they have long decay lifetimes and are

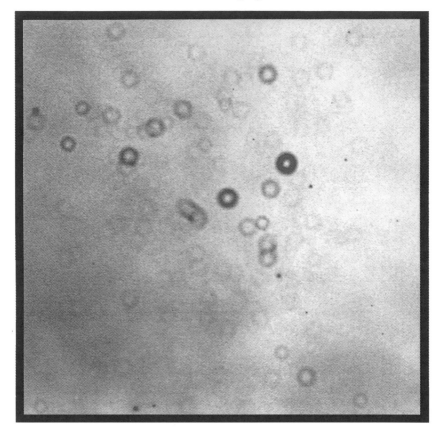

Fig. 4.4. Shown is a typical CCD flat field image. This is an R-band flat field image for a 1,024 × 1,024 Loral CCD. The numerous "doughnuts" are out of focus dust specks present on the dewar window and the filter. The varying brightness level and structures are common in flat field images. As seen in the histogram of this image (Figure 4.1) this flat field has a mean level near 6,950 ADU, with an approximate dispersion of (FWHM) 400 ADU.

very narrow spectrally. In addition, owing to upper atmosphere motions, OH concentrations, and their long decay times, these emission lines are highly variable in time and strength, even within a given night. Dealing with fringes that occur in CCD data can be a difficult problem but one for which a number of solutions exist (Wagner, 1992; Broadfoot & Kendall, 1968).

Additionally, cosmetic effects such as bad pixels, hot pixels (LEDs), or dead columns will be present and can mar a CCD image. Not only do these flaws spoil the beauty of the two-dimensional data, they can cause problems during calibration and analysis by hindering software processes

Fig. 4.5. Shown is a typical CCD object frame showing a star field. This image has been properly reduced using bias frame subtraction and division by a flat field image. Note how the background is of a uniform level and distribution; all pixel-to-pixel nonuniformities have been removed in the reduction process. The stars are shown as black in this image and represent R magnitudes of 15th (brightest) to 20th (faintest). The histogram shown in the remainder of the figure is typical for a CCD object frame after reduction. The large grouping of output values on the left (values less than about 125 ADU) are an approximate Gaussian distribution of the background sky. The remaining histogram values (up to 1,500 ADU) are the pixels that contain signal levels above the background (i.e., the pixels within the stars themselves!).

and not allowing correct flux estimates to be made for the pixels that they affect. Procedures for the removal of or correction for these types of problems can be applied during image calibration and reduction. They are highly specialized tasks, depend on the desired output science goals, and generally are specific to a particular CCD, instrument, or type of observation being made. A complete discussion of all of these topics lies

Fig. 4.5. (*Continued*)

beyond our space limitations but the interested reader will find discussions of such corrections in Gilliland (1992), Gullixson (1992), Djorgovski (1984), Janesick et al. (1987a), and numerous specific instrument manuals and reference papers concerning the finer points of specific CCD related issues (see Appendix A).

4.6.2 Wide-Field CCD Imaging

With the advent of large-footprint CCDs and the construction of CCD mosaic arrays containing many chips, wide-field imaging is becoming one of the major applications in astronomy today. The Wide Field Camera (WFC) on the 2.5-m Isaac Newton Telescope (INT) consists of four 2K × 4K EEV CCDs plus an additional 2K × 2K Loral CCD (operating in frame transfer mode for autoguiding), mounted at prime focus. The pixel size for each CCD is 15 microns and a ten minute exposure through a broad-band filter yields a faint limit of near 25th magnitude. Figure 4.6 presents a view of the INT Wide Field Camera instrument.

By far the most ambitious wide-field imaging project is the MEGA-CAM instrument scheduled to begin operation in 2001 on the CFHT

Fig. 4.6. Photograph of the Wide Field Camera mosaic imager. The mosaic bottom is the old version, which used four Loral 2K × 2K CCDs, while the mosaic top is the current version consisting of four EEV CCDs. In both mosaics, the 2K × 2K CCD off to the side is used in frame transfer mode as an autoguider.

(Boulade et al., 1998). This imager will contain 36 to 40 2K × 4.5K EEV CCDs, each with 13.5-micron pixels. The device will image a full 1 square degree field of view at 0.185 arcsec/pixel resolution. As with all shuttered large-area imagers, MEGACAM will use a half-disk spinning shutter that rotates in a constant direction to ensure uniform exposure time across the array. Current plans call for CCD readout at an amazingly fast 20 seconds per chip.

Wide-field CCD mosaic imagers allow a tremendous amount of information (and data) to be collected in one exposure. The WFC, for example, produces over 800 MB per one mosaic exposure. CCD mosaic arrays are pioneering new scientific advances, driving astronomical technology and bringing us back to many projects started years ago with photographic plates, but now pushing them to new depths and precision levels. The efficiency of a large-area survey can be estimated by the quantity

$$\epsilon = \Omega D^2 q,$$

where Ω is the solid angle of the field of view, D is the diameter of the telescope, and q is the total throughput quantum efficiency of the instrument assuming that the seeing disk is resolved. One can see that the time needed for completion of a survey to a given brightness limit depends inversely on ϵ.

Using wide-field CCD imagers leads to the inevitable result that new issues of calibration and data reduction must be developed. For example, when the field of view of a large-area CCD (whether a single large CCD with a wide field of view or an array of chips) approaches ∼0.5 degrees in size, differential refraction of the images across the field of view of the CCD begins to become important. Color terms therefore propagate across the CCD image and must be corrected for to properly determine and obtain correct photometric information contained in the data.

CCD observations that occur through uniform thin clouds or differential measures essentially independent of clouds are often assumed to be valid, as it is believed that clouds are grey absorbers and that any cloud cover that is present will cover the entire CCD frame. Thus flux corrections (from say previous photometric images of the same field) can be applied to the nonphotometric CCD data, making it usable. Large-field CCD imaging cannot make such claims. A one or more square degree field of view has a high potential of not being uniformly covered by clouds, leading to unknown flux variations across the CCD image.

Observations of large spatial areas using CCD mosaics also necessitate greater effort and expense in producing larger filters, larger dewar windows, and larger correction optics. Variations of the quality and color dependence of these optical components over the entire field of view are generally noticeable. These optical aberrations will cause point spread function (PSF) changes over the large areas imaged with wide-field CCDs. The use of large-format CCDs or CCD mosaics on Schmidt telescopes is increasing and such an imager provides a good example of the type of PSF changes that occur across the field of view (see Figure 4.7). Coma and chromatic aberrations are easily seen upon detailed inspection of the PSFs, especially near the corners or for very red or blue objects whose peak flux lies outside of the color range for which the optics were designed. Thus, for wide-field applications, such as that represented in Figure 4.7, the typical assumption that all the PSFs will be identical at all locations within the field of view must be abandoned.

A more subtle effect to deal with in wide-field imaging is that of the changing image scale between images taken of astronomical objects and those obtained for calibration purposes. For example, a dome flat field image taken for calibration purposes will not have exactly the same image scale per pixel over the entire CCD image as an object frame taken with the same CCD camera, but of an astronomical scene. Since the image scale (arcsec per pixel) changes over large (astronomical) fields of view, flat field and other calibration corrections become complex as the different types of images have different or changing pixel scales. Therefore, during the reduction process, some type of image interpolation or movement of flux from one pixel scale to another must be performed. This is never a desirable procedure, but for wide-field CCD imaging, flat field and other calibration procedures make it necessary at some level. One way around such an issue is to use sky flats, which will have the same pixel scale if obtained at the same zenith distance as your object frame, but they will also have all of the disadvantages of sky flats discussed above in Section 4.2.

As with previous new advances in CCD imaging, wide-field imaging has issues that must be ironed out. However, this exciting new field of research is still in its infancy and those of you reading this book who are involved in such work are the ones who must help determine the proper data collection and reduction procedures to use.

4.6.3 CCD Drift Scanning and Time-Delay Integration

The standard method of CCD imaging is to point the telescope at a particular place in the sky, track the telescope at the sidereal rate, and integrate with the detector for a specified amount of time. Once the

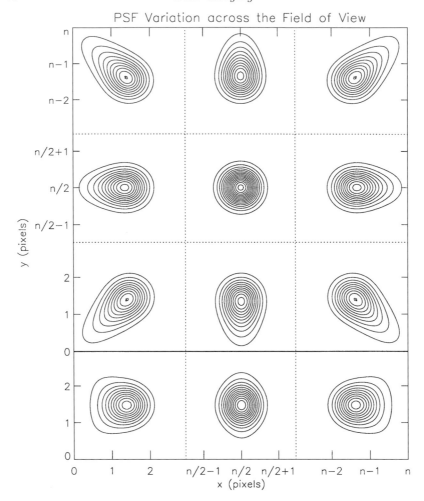

Fig. 4.7. PSF variations for a star imaged at nine locations within the field of view of a large mosaic CCD camera placed at the focal plane of a Schmidt telescope. Only the center of the field has a circular, symmetric PSF while the other positions show extended tails due to optical abberations and chromatic effects. The three PSFs at the bottom of the figure are column sums of the PSFs vertically above them. From Howell et al. (1996).

desired integration time is obtained, the shutter is closed and the CCD is read out. For telescopes incapable of tracking on the sky or to obtain large areal sky coverage without the need for complex CCD mosaics, the techniques of CCD drift scanning and time-delay integration were developed (McGraw, Angel, & Sargent, 1980; Wright & Mackay, 1981).

Drift scanning consists of reading the exposed CCD at a slow rate while simultaneously mechanically moving the CCD itself to avoid image smear. The readout rate and mechanical movement are chosen to provide the desired exposure time. Each imaged object is thus sampled by every pixel in the column thereby being detected with the mean efficiency of all pixels in the column. Nonuniformities between the pixels in a given column are thus eliminated as each final pixel is, in essence, a sum of many short integrations at each pixel within the column. Cross column efficiency differences are still present but the final image can now be corrected with a one-dimensional flat field. Drift scanning also has the additional advantage of providing an ideal color match to background noise contributions, unavailable with dome flats. Very good flat fielding of a traditional image might reach 0.5 or so percent, while a good drift scanned CCD image can be flattened to near 0.1 percent or better (Tyson & Seitzer, 1988). Drift scanning has even been accomplished with IR arrays (Gorjian, Wright, & Mclean, 1997).

Time-delay integration or TDI is a variant on the drift scanning technique. In TDI, the CCD does not move at all but is read out at exactly the sidereal rate. This type of CCD imaging is necessary if astronomical telescopes such as transit instruments (McGraw, Angel, & Sargent, 1980) or liquid mirror telescopes (Gibson, 1991) are to be used. The same flat fielding advantages apply here as in drift scanning but the integration time per object is limited by the size of the CCD (i.e., the time it takes an object to cross the CCD field of view). For a 2,048 × 2,048 CCD with 0.7 arcsec pixels, the integration time would be only 96 seconds at the celestial equator. Rescanning the same area could be performed and co-added to previous scans as a method of increasing the exposure time, but time sampling suffers.

TDI is mechanically simple, as nothing moves but the electrons in the CCD. This charge movement has been termed electro-optical tracking. Large sky regions can be surveyed, albeit to shallow magnitude limits, very quickly using TDI. Overhead time costs for TDI only consist of the "ramp up" time, that is, the time needed for the first objects to cross the entire field, and the scan time. Using our same 2048 × 2048 CCD as in the example above, we find that a 23 arcsec by 3 degree long strip of the sky at the celestial equator can be scanned in about 2–3 minutes compared with the nearly 25 minutes required if pointed observations of equivalent integration are used.

Although drift scanning and TDI are seemingly great solutions to flat fielding issues and offer the collection of large datasets, drift scanning requires the CCD to move during the integration with very precise and repeatable steps. This is quite a mechanical challenge and will increase the cost of such an instrument over that of a simple CCD imager. In addition, both techniques suffer two potential drawbacks (Gibson & Hickson, 1992). Images obtained by drift scanning and TDI techniques have elongated PSFs in the east–west direction. This is due to the fact that the rows of the CCD are shifted discretely while the actual image movement is continuous. We note here that objects seperated by even small declination differences (i.e., one CCD field of view) do not have the same rate of motion. The resulting images are elongated east–west and are a convolution of the seeing with the CCD pixel sampling.

TDI imagery contains an additional distortion in the north–south direction owing to the cuvature of an object's path across the face of the CCD (if imaging away from the celestial equator). This type of distortion is usually avoided in drift scan applications as the telescope and CCD tracking are designed to eliminate this image smearing. This sort of mechanical correction can not be applied to TDI imaging.

These image deformations have been studied in detail (Gibson & Hickson, 1992) and are seen to increase in magnitude for larger format CCDs or declinations farther from the celestial equator. For example, at a declination of ±30 degrees, a 1 arcsec per pixel CCD will show an image smear of about 6 pixels. One solution to this large image smear is to continuously reorient the CCD through rotations and translations, such that imaging scans are conducted along great circles on the sky rather than a polar circle or at constant declination. Such a mechanically complex device has been built and used for drift scanning on the 1-m Las Campanas telescope (Zaritsky, Shectman, & Bredthauer, 1996). Another solution is the development of a multilens optical corrector that compensates for the image distortions by tilting and decentering the component lenses (Hickson & Richardson, 1998).

A few telescopes have made good use of the technique of drift scanning or TDI, providing very good astronomical results. Probably the first such project was the Spacewatch telescope (Gehrels et al., 1986) setup to discover and provide astrometry for small bodies within the solar system. Other notable examples are the 2-m transit telescope previously operated on Kitt Peak (McGraw, Angel, & Sargent, 1980) and a 2.7-m liquid mirror telescope currently running at the University of British Columbia (Hickson et al., 1994). This latter telescope contains a rotating mercury mirror and images a 21 arcminute strip of the zenith with an effective integration time of 130 seconds. Using TDI, a typical integration with this liquid mirror telescope reaches near 21st magnitude in R and continuous readout of the CCD produces about 2 GB of data per night.

Two present-day examples of telescopes employing drift scanning and TDI techniques are the QUEST telescope (Sabby, Coppi, & Oemler, 1998) and the Sloan digital sky survey (Gunn et al., 1998). The QUasar Equatorial Survey Team (QUEST) telescope is a 1-m Schmidt telescope that will provide UBV photometry of nearly 4,000 square degrees of the sky to a limiting magnitude of near 19. The focal plane will contain 16 2,048 × 2,048 Loral CCDs arranged in a 4 × 4 array. The telescope is parked and the CCDs are positioned such that the clocking (column) direction is east–west and the readout occurs at the apparent sidereal rate. Each object imaged passes across four CCDs covered, in turn, with a broad-band V, U, B, and V filter. The effective integration time (i.e., crossing time) is 140 seconds, providing nearly simultaneous photometry in U, B, and V.

As we have seen above, a problem with drift scanning is that the paths of objects that drift across the imager are not straight and they can cross the wide field of view with different drift rates. We have discussed a few solutions to these issues, and in the QUEST project (Sabby, Coppi, & Oemler, 1998) we find another. The CCDs are fixed, in groups of four, to long pads lying in the north south direction. These pads can pivot independently such that they align perpendicular to the direction of the stellar paths. The CCDs are also able to be clocked at different rates, with each being read out at the apparent sidereal rate appropriate for its declination.

The Sloan digital sky survey (SDSS) is a large-format mosaic CCD camera consisting of a photometric array of 30 2,048 × 2,048 SITe CCDs and an astrometric array of 24 400 × 2,048 CCDs (see Figure 4.8). The photometric CCDs are arranged in six columns of five CCDs each, providing essentially simultaneous five-color photometry of each image object. The astrometric CCDs are mounted in the focal plane above and below the main array and will be used to provide precise positional information needed for the follow-up multifiber spectroscopy. The SDSS will use a 2.5-m telescope located in New Mexico to image one quarter of the entire sky down to a limiting magnitude of near 23.

TDI scans along great circles will be used by the SDSS to image a region of the sky 2.5 degrees wide. Using six intermediate band filters, covering 3,550 Å to 9,130 Å, scanning at the sidereal rate provides an effective integration per color of 54 seconds with a time delay of 72 seconds

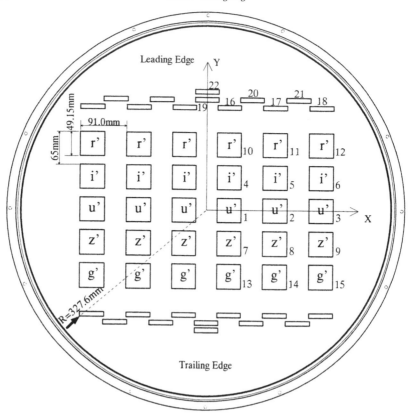

Fig. 4.8. The optical layout within the dewar of the Sloan digital sky survey CCD imager. The right side of the figure labels the CCDs as to their function; 1–15 are photometric CCDs, 16–21 are astrometric CCDs, and 22 (top and bottom) are focus CCDs. The left side gives the dimensions of the array. The labels r'–g' denote the five seperate intermediate band filters, each a single piece of glass covering all six horizontal CCDs. The scan direction is upward causing objects to traverse the array from top to bottom. From Gunn et al. (1998).

between colors caused by CCD crossing time and chip spacing. Complete details of the SDSS, too lengthly for presentation here, can be found in Gunn et al. (1998).

The SDSS has just begun operation and is probably the largest, most complex camera of its kind. Other projects of a similar nature are discussed in Boulade et al. (1998), Gunn et al. (1998), and Miyazaki et al. (1998).

5

Photometry and Astrometry

One of the basic astronomical pursuits throughout history has been to determine the amount and temporal nature of the flux emitted by an object as a function of wavelength. This process, termed photometry, forms one of the fundamental branches of astronomy. Photometry is important for all types of objects from planets to stars to galaxies, each with their own intricacies, procedures, and problems. At times, we may be interested in only a single measurement of the flux of some object, while at other times we could want to obtain temporal measurements on time scales from seconds or less to years or longer.

We start this chapter with a brief discussion of the basic methods of performing photometry when using digital data from 2-D arrays. It will be assumed here that the CCD images being operated on have already been reduced and calibrated as described in detail in the previous chapter. We will see that photometric measurements require that we accomplish only a few steps to provide output flux values. Additional steps are then required to produce light curves or absolute fluxes.

Some photometric output products, such as differential photometry, require fewer additional steps, whereas to obtain the absolute flux for an object, additional CCD frames of photometric standards are needed. These standard star frames are used to correct for the Earth's atmosphere, color terms, and other possible sources of extinction that may be peculiar to a given observing site or a certain time of year (Pecker, 1970).

As an introduction to the level of atmospheric extinction one might expect as a function of observational elevation and wavelength, Table 5.1 lists values of the extinction in magnitudes resulting from the Earth's atmosphere for an observing site at 2,200 m elevation. Note that for observations made at reasonable airmass and redward of 4,000 Å, the

75

Table 5.1. *Example atmospheric extinction values (magnitudes)*

Altitude	Airmass	3000 Å	3500 Å	4000 Å	4500 Å	5000 Å	5500 Å	6000 Å	6500 Å	7000 Å	8000 Å	9000 Å	10000 Å
90	1.00	1.2	0.65	0.4	0.3	0.2	0.2	0.2	0.1	0.1	0.1	0.1	0.1
75	1.04	1.2	0.65	0.4	0.3	0.2	0.2	0.2	0.1	0.1	0.1	0.1	0.1
60	1.15	1.3	0.75	0.5	0.3	0.3	0.2	0.2	0.2	0.1	0.1	0.1	0.1
45	1.41	1.6	0.9	0.6	0.4	0.3	0.3	0.2	0.2	0.2	0.1	0.1	0.1
30	1.99	2.3	1.3	0.8	0.6	0.4	0.4	0.3	0.3	0.2	0.2	0.1	0.1
20	2.90	3.3	1.55	1.2	0.8	0.6	0.5	0.5	0.4	0.3	0.2	0.2	0.2
15	3.82	4.4	2.5	1.6	1.1	0.8	0.7	0.6	0.5	0.4	0.3	0.3	0.2
10	5.60	6.4	3.65	2.3	1.6	1.2	1.1	1.0	0.7	0.6	0.4	0.4	0.3
5	10.21	11.8	6.7	4.2	2.9	2.2	1.9	1.7	1.4	1.1	0.8	0.7	0.6

effect of the Earth's atmosphere is, at worst, a few tenths of a magnitude. The details of photometric corrections for the Earth's atmosphere and extinction effects are not germane to the topic of this book and their discussion here would be beyond the allowed space limitations. The interested reader is referred to the excellant presentations in Dacosta (1992), Hendon & Kaitchuck (1982), and Young (1974). Further good discussions of photometric data handling are presented in Merline & Howell (1995), Stetson (1987), Howell (1992), Hiltner (1962), Walker (1990), Kreidl (1993), Howell & Jacoby (1986), and Howell, Everett, & Ousley (1999).

5.1 Stellar Photometry from Digital Images

Prior to the time when CCDs became generally available to the astronomical community, digital images of astronomical objects were being produced by detectors such as silicon intensified targets (SITs), video tube-type cameras, image tubes, and electronographic cameras. In addition, scanning of photographic plates with a microdensitometer resulted in large amounts of digital output. These mechanisms produced digital data in quantity and at rates far in excess of the ability of workers to individually examine each object of interest within each image. Today, the amount of CCD data greatly exceeds this limit. Thus, in the early 1980s, work began in earnest to develop methods by which photometry could be obtained from digital images in a robust, mostly automated manner.

One of the first such software packages to deal with digital images was written by Adams, Christian, Mould, Stryker, and Tody (Adams et al., 1980) in 1980. Numerous other papers and guides have been produced over the years containing methods, ideas, entire software packages that perform photometry, and specific detailed information for certain types of objects. I have tried to collect a fairly complete list of these in Appendix A. While details vary, the basic photometric toolbox must contain methods that perform at least the following primary tasks: (*i*) image centering, (*ii*) estimation of the background (sky) level, and (*iii*) calculation of the flux contained within the object of interest. We will assume below, for simplicity, that we are working with stellar images that to a good approximation are well represented by a point spread function of more-or-less Gaussian shape. Deviations from this idealistic assumption and nonpoint source photometry will be discussed as they arise.

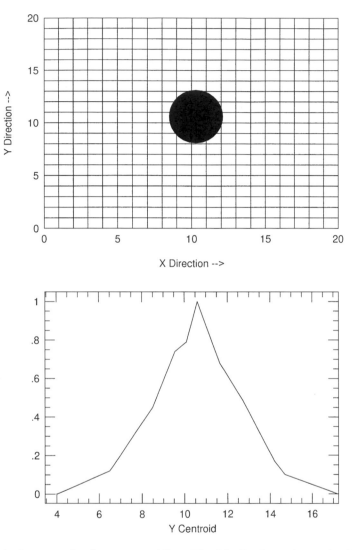

Fig. 5.1. An example of x, y centroiding. The idealized star image in the top box sits on a pixel grid with a center of $(x, y) = (10.3, 10.6)$. The two other plots represent the x, y centroids for the star image normalized to a maximum height of one. In this case, the star center is well approximated by the peaks in the x, y centroids.

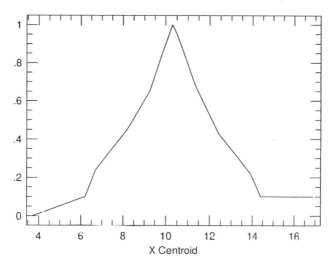

Fig. 5.1. (*Continued*)

5.1.1 Image Centering

Probably the simplest and most widely used centering approximation for a point spread function (PSF) is that of marginal sums or first moment distributions. Starting with a rough pointer to the position of the center of the star (e.g., the cursor position, reading off the x, y coordinates, or even a good guess), the intensity values of each pixel within a small box centered on the image and of size $L \times L$ (where L is comparable to the size of the PSF) are summed in both x and y directions (see Figure 5.1). The x, y center is computed as follows: The marginal distributions of the PSF are found from

$$I_i = \sum_{j=-L}^{j=L} I_{i,j}$$

and

$$J_j = \sum_{i=-L}^{i=L} I_{i,j},$$

where $I_{i,j}$ is the intensity (in ADU) at each x, y pixel; the mean intensities are determined from

$$\bar{I} = \frac{1}{2L+1} \sum_{i=-L}^{i=L} I_i$$

and

$$\bar{J} = \frac{1}{2L+1} \sum_{j=-L}^{j=L} J_j;$$

and finally the intensity weighted centroid is determined using

$$x_c = \frac{\sum_{i=-L}^{i=L}(I_i - \bar{I})x_i}{\sum_{i=-L}^{i=L}(I_i - \bar{I})}$$

for all $I_i - \bar{I} > 0$ and

$$y_c = \frac{\sum_{j=-L}^{j=L}(J_j - \bar{J})y_j}{\sum_{j=-L}^{j=L}(J_j - \bar{J})}$$

for all $J_j - \bar{J} > 0$.

For well-sampled (see Section 5.6), relatively good S/N (see Section 4.4) images, simple x, y centroiding provides a very good determination of the center position of a PSF, possibly as good as 1/5th of a pixel. More sophisticated schemes to provide better estimations of image centers or applications appropriate to various types of non-Gaussian PSFs are given in Davis (1994), Massey & Davis (1992), Lasker et al. (1990b), Stone (1989), Penny & Dickens (1986), Chiu (1977), and Howell et al. (1996).

5.1.2 Estimation of Background

The importance of properly estimating the background level on a CCD resides in the fact that the same pixels that collect photons of interest from an astronomical source also collect photons from the "sky" or background, which are of no interest. Remember that the background or sky value in a CCD image contains not only actual photons from the sky but also photons from unresolved astronomical objects, read noise, thermally generated dark current electrons, and other sources. All of these unwanted additional photons must be accounted for in some manner, estimated, and removed from the image before a final determination of the source flux is made. In order to determine this background level, a common technique is to place a software annulus around the source of interest and then use statistical analysis to estimate its mean level on a per pixel basis.

The background or sky annulus is usually defined by an inner and outer radius or by an inner radius and a width (see Figure 5.2). One simple, yet powerful, manner by which an estimation of the background

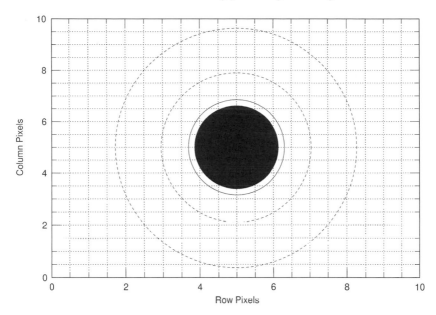

Fig. 5.2. Schematic drawing of a stellar image on a CCD pixel grid. The figure shows the location of the star, the "star" aperture (solid line), and the inner and outer "sky" annuli (dashed circles).

level can be made is to simply extract the values of all the pixels within the annulus, sum them, and divide by the total number of pixels within the annulus. This provides an average value per pixel for the background level of the CCD image. For a good statistical determination of the background level, the total number of pixels contained within this annulus should be relatively large, about three times the number contained within that of the source aperture (Merline & Howell, 1995).[†] A more robust estimator, requiring very little additional work, is to collect all the pixel values from inside the sky annulus, order them in increasing value, and find the median intensity, B_M.[‡] A nice touch here is to reexamine

[†] Partial pixels arising from placing a circular annulus on a rectangular grid are usually not of concern here, as the number of annulus pixels is large. However, partial pixels cannot be so easily dismissed when we are determining the intensity within the much smaller source aperture.

[‡] Note that the statistical values determined from a CCD image for the median or the mode must pick their answer from the list of the actual CCD pixel ADU values, that is, values that are integers containing only digitized levels and thus digitization noise. The statistical mean, however, allows for noninteger answers. This seemingly subtle comment is of great importance when dealing with partial pixels, undersampled data, or high CCD gain values.

the list of annulus pixel values and toss out all those with values greater than $\pm 3\sigma$ from B_M. This last step will eliminate cosmic ray hits, bad pixels, and contamination from close by astronomical neighbors if they exist.

When applying a median filter and the 3σ cutoff technique to the list of background pixels, one can use the remaining annulus pixel values to construct a background histogram computed with a bin width resolution of say 0.5 ADU (Figure 5.3). The background histogram will be centered on the median background value with all contained pixel values within $\pm 3\sigma$. Since the detailed resolution of 0.5 ADU binning

Fig. 5.3. Histogram of the "sky" annulus around a star in the CCD image shown in Figure 4.5. Notice the roughly Gaussian shape to the sky distribution but with an extended tail toward larger values. This tail is due to pixels that were not completely calibrated in the reduction process, pixels with possible contamination owing to dark current or cosmic rays, pixels with increased counts owing to unresolved PSF wings from nearby stars, and contamination of sky annulus pixels by faint unresolved background objects. The need for some histogram smoothing, such as that described in the text, is apparent, especially near the peak of the distribution.

will likely produce a ragged histogram (since only a finite number of background pixels are used), some smoothing of the histogram may be useful, such as that done by Lucy (1975). Lucy smoothing will broaden the histogram distribution slightly after one iteration but application of a second iteration will restore the correct shape and provide a smooth histogram from which to proceed. This final step produces a statistically valid and robust estimator from which we can now compute the mean value of the background, $\bar{B}$. A determination of the centroid of the final smoothed histogram, using equational forms for the centroid in one dimension such as those discussed above, can now be applied. Centroiding of this smoothed histogram is not influenced by asymmetries that may have been present in the wings of the initially unsmoothed values.

Correct estimation of the level of the CCD background on a per pixel basis is of increasing importance as the S/N of the data decreases and/or the CCD pixel sampling becomes poor. A background level estimation for each pixel that is off by as little as a few ADU can have large effects on the final result (Howell, 1989). Determination of the CCD background level has an associated error term $\propto (1 + n_{\text{pix}}/n_B)^{-1/2}$, which should be included in the S/N calculation of the final result (see Section 4.1).

5.1.3 Estimation of Point Source Intensity

We now come to the pixel values of interest, that is, those which contain photons from the source itself. Using a software aperture of radius r centered on the x, y position of the centroid of the source PSF, we can extract the values from all pixels within the area A ($= \pi r^2$) and sum them to form the quantity S, the total integrated photometric source signal. The sum S contains contributions from the source but also from the underlying background sources within A as well. To remove the estimated contribution to S from the background, we can make use of the value $\bar{B}$, discussed above. We can calculate an estimate of the collected source intensity, I, as $I = S - n_{\text{pix}}\bar{B}$, where n_{pix} is the total number of pixels contained within the area A. There are some additional minor considerations concerning this procedure but these will not be discussed here (Merline & Howell, 1995).

A final step usually performed on the quantity I, which we will discuss further below, is to determine a source magnitude. The value of a magnitude is defined by the following standard equation:

$$\text{Magnitude} = -2.5 \log_{10}(I) + C,$$

where I is the source intensity per unit time, that is, the flux (universally given as per second), and C is an appropriate constant (usually $\sim$23.5–26 for most earthly observing sites) and determined in such a manner so that the calculated source magnitude is placed on a standard magnitude scale such as that of the Johnson system or the Strömgren system.

As we mentioned above, when using circular apertures on rectangular pixel grids, partial pixels are inevitable. While we could toss them away for the large background area, we cannot follow a similar sloppy procedure for the smaller source aperture. Thus the question becomes, how do we handle partial pixels? This is not a simple question to answer and each photometric software package has its own methodology and approach. The three choices a software package can use are:

1. Do not use partial pixels at all. Any source intensity that falls into the source aperture but within a partially inscribed pixel is simply not used in the calculation of S.
2. Sum the values for every pixel within the source aperture regardless of how much or how little of the pixel area actually lies within A.
3. Make use of some computational weighting scheme that decides, in a predefined manner, how to deal with the counts contained within each partial pixel in the source aperture.

This last choice often uses the ratio of the pixel area inside the source aperture to that outside the aperture as a simple weighting factor. A computational scheme to handle partial pixels in a software package designed to perform digital photometry is the hardest of the above choices to implement, but it will provide the best overall final results. To know exactly how a certain software package handles partial pixels, the user is referred to the details presented within the documentation provided with the software. Many PC-type packages that perform photometry on CCD images do not detail their partial pixel and magnitude calculation methods and are therefore "black boxes" to be avoided.

There are two basic methods by which most observers estimate the total integrated signal within their source aperture: point spread function fitting and digital aperture photometry. The first method relies on fitting a 2-D function to the observed PSF and using the integrated value underneath this fitted function as an estimate of S. The second method, digital aperture photometry, attempts to place a software aperture about the source profile (as shown in Figure 5.2), centered in some manner (e.g., x, y centroids), and then simply sums the pixel values within the source aperture to provide the estimation of S. We will discuss each of these methods in turn below and note here that it is unlikely that a single method of estimation will be the best to use in all possible situations. For example, for severely undersampled data the method of pixel mask fitting (Howell et al., 1996) provides the best solution.

5.2 Two-Dimensional Profile Fitting

The profiles of astronomical point sources that are imaged on two-dimensional arrays are commonly referred to as point spread functions or PSFs. In order to perform measurements on such images, one method of attack is profile fitting. PSFs can be modeled by a number of mathematical functions, the most common include Gaussian,

$$G(r) \propto e^{\left(\frac{r^2}{2a^2}\right)},$$

modified Lorentzian,

$$L(r) \propto \frac{1}{1 + (r^2/a^2)^b},$$

and Moffat,

$$M(r) \propto \frac{1}{(1 + r^2/a^2)^b},$$

representations, where r is the radius of the point source and a and b are fitting parameters (Stetson, Davis, & Crabtree, 1990). These types of functional forms can be used to define the PSF for each star within an image by the assumption that they provide a good representation of the data itself. For example, adjustment of the values of a and b within one of these functions may allow an imaged PSF to be matched well in radius and profile shape (height and width), allowing a simple integration to be performed to measure the underlying flux.

Generally, the above functions are only a fair match to actual PSFs and so a second method of profile fitting can be applied. This method consists of using an empirical PSF fit to the actual digital data itself, producing modified versions of the above functions. Depending on the application, PSFs may be evaluated at the center of a pixel or integrated over the area of each pixel. Even more general methods of allowing the data to produce completely analytic forms for the PSF functions have been attempted. The techniques and use of empirical PSFs could fill an entire chapter; we refer the reader to Stetson (1987), Diego (1985), and King (1971) for more details.

Fig. 5.4. Stellar PSFs are shown for various cases. The figure above shows two model PSFs, one for a bright star (S/N ∼ 125) and one for a faint star (S/N ∼ 20). The remaining two panels show similar brightness stars but are actual CCD data. Note that the models are shown as 3-D pixel histograms whereas the real data are represented as spline profile fits to the actual PSFs. The disadvantage of the latter type of plotting is that the true pixel nature of the image is lost.

Both techniques, the use of completely mathematical forms for a PSF approximation and the more empirical method, have their advantages and disadvantages. Model PSF fitting allows the necessary integrations and pixel interpolations to be carried out easily as the functions are well known, while the empirical method, which makes hardly any

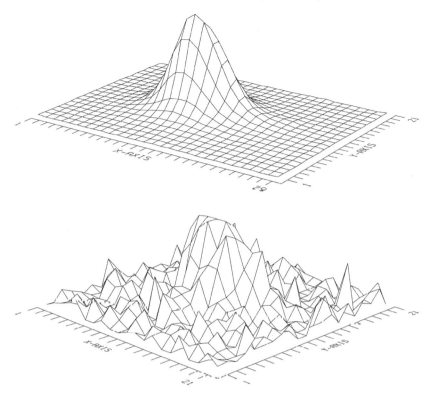

Fig 5.4. (*Continued*)

assumptions about the actual shape of the PSF, is only defined on the CCD pixel grid and not in any general mathematical way. This latter complication can cause difficulties when trying to interpolate the empirical shape of one PSF somewhere on the CCD (say a bright reference star) to a PSF located somewhere else on the same image but that is likely to have a different pixel registration. For this reason, some implementations of empirical PSF fitting actually make use of the sum of an analytic function (such as one of those given as above) and a look-up table of residuals between the actual PSF and the fitting function. Figure 5.4 shows examples of some PSF models and some actual PSFs obtained with CCDs.

Procedurally, profile fitting techniques work by matching the implied PSF to the actual digital data in a 2-D fashion and within some radius, r, called the fitting radius. An attempt is then made to maximize some goodness-of-fit criteria between the assumed PSF and the observed one.

PSF fitting can be further optimized by fitting N point sources within the image simultaneously (usually one uses the brightest stars within the image) and using some combination of statistically weighted mean values for the final fitting parameters. PSF fitting can be very computationally demanding, much more so than the method of aperture photometry discussed below. However, for some types of imagery, for example crowded fields such as within star clusters for which some PSFs may overlap, PSF fitting may be the only method capable of producing scientifically valid results (Stetson, 1987).

5.3 Aperture Photometry

Aperture photometry is a technique that makes no assumption about the actual shape of the source PSF but simply collects and sums up the observed counts within a specified aperture centered on the source. The aperture used may be circular (usually the case for point sources), square, or any shape deemed useful. Aperture photometry is a simple technique, both computationally and conceptually, but this same simplicity may lead to errors if applied in an unsuitable manner or when profile fitting is more appropriate.

The basic application of aperture photometry starts with an estimate of the center of the PSF and then inscribes a circular software aperture of radius r about that center. The radius r may simply be taken as three times the full-width at half-maximum ($r = 3 \cdot \text{FWHM}$): the radius of a PSF that would contain 100% of the flux from the object (Figure 5.5) (Merline & Howell, 1995). Summing the counts collected by the CCD for all the pixels within the area $A = \pi r^2$, and removing the estimated background sky contribution within A, one finally arrives at an estimated value for I. We see again that partial pixels (a circular software aperture placed on a rectangular grid) are an issue, even for this simple technique. Using a square or rectangular aperture alleviates the need for involving partial pixels but may not provide the best estimate of the source flux. Noncircular apertures do not provide a good match to the 2-D areal footprint of a point source, thereby increasing the value of n_{pix} that must be used and decreasing the overall S/N of the measurement. Remember, however, that for bright sources, n_{pix} is essentially of no concern (see Section 4.4).

It has been shown (Howell, 1989; Howell, 1992) that there is a well-behaved relation between the radius of the aperture of extraction of a point source and the resultant S/N obtained for such a measurement.

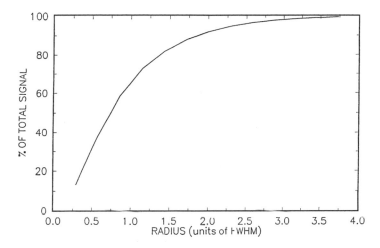

Fig. 5.5. For any reasonable PSF approximation, the figure above shows the run of the total encircled signal with radius of the PSF in FWHM units. Note that within a radius of $3 \cdot$ FWHM essentially 100% of the signal is included.

An optimum radius aperture, that is, one that provides the optimum or best S/N for the measurement, can be determined for any given PSF and generally has a radius of near 1 FWHM. This optimum radius is a weak function of the source brightness, becoming smaller in size for fainter sources. Figure 5.6 illustrates this idea for three point sources of different brightness.

To understand the idea of an optimum radius and why such a radius should exist, one simply has to examine the S/N equation given in Section 4.4 in some detail. To obtain a higher S/N for a given measurement, more signal needs to be collected. To collect more signal, one can use a larger aperture radius, up to the maximum of $3 \cdot$ FWHM. However, the larger r is, the more pixels that get included within the source aperture and the larger the value of n_{pix}. As n_{pix} increases, so does the contribution to the error term from noise sources other than the source itself. Thus, a balance between inclusion of more signal (larger r) and minimizing n_{pix} in the source aperture (smaller r) leads to an optimum extraction radius for a given source.

We saw in Figure 5.5 that if extracted at or very near an aperture radius of $3 \cdot$ FWHM, 100% of the light from a point source would be collected. However, to obtain the maximum S/N from your measurement, extraction at a smaller radius is warranted. If one extracts the source

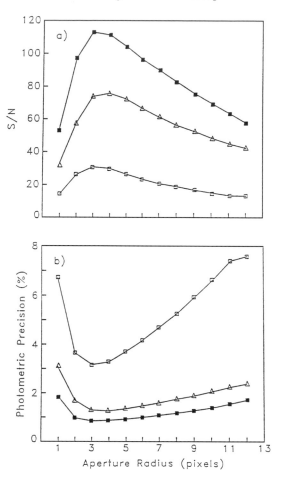

Fig. 5.6. The S/N obtained for the measurement of a point source is not constant as a function of radius. There is an optimum radius at which the S/N will be a maximum. The top panel shows this effect for three point sources that differ in brightness by 0.3 (middle curve) and 2.0 (bottom curve) magnitudes compared with the top curve (filled squares). The bottom panel presents the same three stars as a function of their photometric precision. The image scale is 0.4 arcsec/pixel and the seeing (FWHM) was near 1.2 arcsec. From Howell (1989).

signal using an aperture that is smaller then the actual PSF radius itself, some of the source light that was collected by the CCD is not included in the summing process and is thus lost. This sounds like an incorrect methodology to use, but remember that inclusion of many pixels lying far from the source center also means inclusion of additional noise contributions to the aperture sum. Therefore, while one may wish to obtain the maximum S/N possible through the use of a smaller aperture (i.e., summation of less than the total collected source counts), for the final result it is often necessary to correct the answer obtained for this shortcoming.

In order to recover the "missing light," one can make use of the process of aperture corrections or growth curves as detailed by Howell (1989) and Stetson (1992). Growth curves do not make any demands on the underlying PSF except through the assumption that any bright stars used to define the aperture corrections are exact PSF replicas of any other (fainter) stars that are to be corrected. As we can see in Figure 5.7, growth curves for the brightest stars follow the same general shape, leading to minor or no necessary aperture corrections at a radius of 3·FHWM. The fainter stars, however, begin to deviate from the canonical growth curve at small radii, resulting in 0.5 up to 1.5 magnitudes of needed aperture correction. In general, as a point source becomes fainter, the wings of the source will contain pixels that have an increasingly larger error contribution from the background, leading to greater deviations from a master growth curve at large r, and thus a larger aperture correction will be needed.

As we will see below, if differential photometric results are desired, the aperture corrections used to bring the extracted source signal back to 100% are not necessary. This is only strictly true if all point sources of interest (those to be used in the differential measures) are extracted with the same (optimum) aperture[†] and have identical PSFs.

5.4 Absolute versus Differential Photometry

Whether an observer needs to obtain absolute or differential photometric measurements depends on the objectives of the scientific and observational program. Absolute photometry results in a measurement of a given astronomical source leading to the true level of the flux received, say in ergs sec^{-1}, or the total luminosity of a source in ergs, each within the specific bandpass used (e.g., a Johnson R filter). Differential photometry

---

[†] Note, it is likely that on a given CCD image all stars of interest will not be of exactly the same brightness and will therefore not all have exactly the same optimum aperture radius (see Figure 5.6). Thus, a compromise is usually needed in which the extraction radius used for all sources of interest is set to that of the optimum size for the faintest stars. This procedure allows the faintest sources to produce their best possible S/N result while decreasing the S/N for bright stars only slightly.

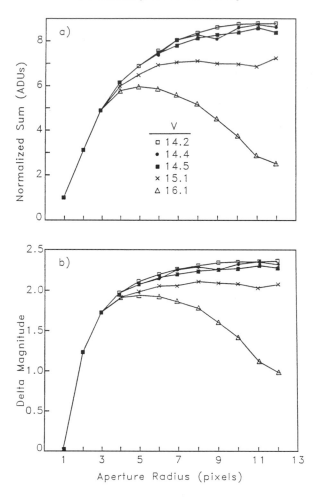

Fig. 5.7. Growth curves for five stars on a single CCD frame. The three bright-est stars follow the same curve, which is very similar to the theoretical expec-tation as shown in Figure 5.5. The two faint stars start out in a similar manner, but eventually the background level is sufficient to overtake their PSF in the wings and they deviate strongly from the other three. Corrections, based on the bright stars, can be applied to these curves to obtain good estimates of their true brightnesses. The top panel presents growth curves as a function of normalized aperture sums while the bottom panel shows the curves as a function of magnitude differences within each successive aperture. The rela-tive magnitudes of the point sources are given in the top panel and the image scale is the same as in Figure 5.6. From Howell (1989).

is just that: The final result is only known with respect to one or more other sources. This is a relative measurement, which, if the other source(s) are known in an absolute manner, can be placed on an absolute scale. All scientific measurements are really differential ones. The difference between absolute and differential measurements is simply that differencing from a known (in an absolute sense) source allows for an absolute result, whereas differencing from an unknown (in an absolute sense) source can only produce a relative result.

In photometric work one can view this in the following way. If the star Vega has its absolute flux known as a function of color (Tüg, White, & Lockwood, 1977), then a comparison of an unknown star with Vega would allow one to calibrate the unknown in an absolute sense. This differential measurement would produce an absolute result. However, if two unknown sources are compared over time, one may be seen to vary periodically allowing a scientific result to be produced, but its absolute flux is unknown. This is a differential measurement that produces only a relative result. If you are concerned about the true brightness of an object, say within a standard photometric system, or its color (e.g., B) in order to place it on an H–R diagram or to obtain its age or metallicity, then absolute photometry is needed. However, if you are after relative variability, say for periodicity analysis, or color indices (e.g., $B - V$) of a large group of stars such as a cluster in which you are looking for outliers, then differential photometry is likely to suffice.

A discussion of the observational and reduction techniques involved in absolute photometry is beyond the scope of this section. The reader is referred to any of the numerous books and articles that discuss the conversion of digital photometric data to (absolute) magnitudes and standard photometric systems (e.g , Dacosta (1992), Hendon & Kaitchuck (1982), Hiltner (1962), and Young (1974)). The observational methodologies and equational transformations used to convert from CCD instrumental magnitudes to standard photometric system magnitudes are essentially identical to those used for similar observational programs with photomultiplier tubes (PMT). Absolute photometry from CCD data has been shown to be just as reliable and every bit as good as PMT measurements (Walker, 1990; Kreidl, 1993; Young, 1974). At the present, typical very good values for errors in the source magnitude for a absolute photometric result from CCD observations are $\pm 1\%$ or less.

Differential photometry concerns itself with the measurement of the difference in brightness of one or more astronomical sources when compared with one or more reference sources. With a two-dimensional CCD array, unlike the situation when using a PMT, it is often the case that multiple sources are imaged simultaneously and can therefore be used as references by which to compare the source(s) of interest. The assumption that is made in this simple approach is that the reference source(s) is not variable (at least over the time period of the observation) and the object(s) of interest can be compared with the reference source(s) on a frame by frame basis. The last step is important as it cancels out any seeing or atmospheric effects that may change with time. Proper statistical comparison of the object of interest and the reference(s) must take into account all sources of error as well as photon statistics in order to use differential photometry correctly (Howell & Jacoby, 1986; Howell, 1992; Howell, Mitchell, & Warnock, 1988; Honeycutt, 1992). Differential techniques allow one to obtain incredible precisions from CCD data. For example, photometric errors of near ± 0.001 magnitude are easily obtainable with differential techniques and even higher precisions, approaching or reaching the theoretical limit, have been realized (Howell, 1993; Gilliland et al., 1993; Howell et al., 1999; Howell, Everett, & Ousley, 1999).

5.5 Astrometry

The science of the exact location and possible movement of astronomical sources has also been greatly advanced through the use of CCDs. Differential astrometric measurements of 1 milliarcsecond (mas) or better are achievable today, yielding new information on positions, proper motions, and parallaxes of astronomical objects. In differential astrometry, one measures the position of an object with respect to other objects within the same CCD field of view. This type of measurement is applicable to point source PSFs (e.g., stars) as well as moving objects (e.g., comets) and fuzzy objects (e.g., galaxies); however, the two latter types of object have additional complexities not discussed here (Monet et al., 1991; Lasker et al., 1990a; Monet, 1992).

Astrometric information is one method by which distance determination may be made for an object of interest. Of course the desire for a distance does not always translate into the ability to obtain such information. In general, in one night of obtaining a few good CCD observations, the determined position for a "well-behaved" star can be known to an astrometric accuracy of about 3 mas. A few tens of CCD frames taken over the course of a year (at a minimum of three different epochs) will allow astrometric accuracies near ± 1 mas to be reached, while the best astrometric positions to date have errors of ± 0.5 mas or less.

Astrometry performed with front-side illuminated CCDs will suffer from the effects of the overlying gate structures in ways such as we have discussed previously; that is, locating the exact center of the PSF is not without bias. The front-side gates must be traversed by incoming photons leading to a decreased (or no) blue response, intrapixel variations in photometric (QE) response (Jorden, Deltorn, & Oates, 1994), and the need to use color terms to adjust the measured PSF centroid position to that of the actual source position. We will discuss an interesting CCD effect caused by intrapixel QE variations in the next chapter.

Back-side illuminated CCDs present a different situation for astrometric work entirely. As mentioned earlier, their response to blue light is much improved and no major source of intrapixel deviation exists. The physical flatness of a back-side illuminated CCD over its surface can be of concern and can introduce an additional term to be corrected for when measuring source positions and relative source offsets. Additionally, thinned CCDs may allow very long wavelength light to pass entirely through the device, be reflected back into the CCD itself, and collected at a slightly different location (e.g., a neighboring pixel) than its original incoming path location. This long wavelength reflection effect can cause a slight blurring of the measured light from the source, a particularly worrisome issue for astrometry of very late type stars. The best astrometric measurements to date, that is, the ones with the smallest rms error and greatest repeatability, have all been made with thinned, back-side illuminated CCDs.

As with any CCD imaging or photometric observation, the use of a particular type of CCD and filter combination will produce a different effective wavelength of the imaged scene and thus a change in the relative positions of different sources as a function of the zenith distance of the observation and the color (spectral type) of each individual source. There is no simple or even complete solution to this issue of differential color refraction but there are some items to note.[†] The use of narrow band filters in front of your CCD is of help in this situation because such filters greatly restrict the bandpass of the observation, thereby reducing the effects of differential refraction and color terms. However, the use of narrow band filters is usually impractical owing to the large loss of incoming signal from the astronomical sources of interest. Astrometric observations made in long wavelength (red) bandpasses have merit as they eliminate much of the refractive effects to start with. Finally, CCD observations obtained near source meridian passage are also a plus given

[†] While reading the remainder of this paragraph, the reader may wish to skip ahead and glance at Table 6.2.

that for an airmass of 1.0, refractive changes are essentially zero within the entire CCD field of view. Details of astrometric observations and a discussion of such effects is presented in Monet et al. (1991) and Monet & Dahn (1983).

Issues of related concern for precision astrometry involve image focus, seeing, use of different filters, PSF stability, telescope collimation, and many others (Monet, 1992). Detailed information concerning one's CCD is vital when attempting precision astrometry. For example, the standard pixel size value available in most of the literature for a 15-micron pixel TI CCD states that the pixel size is 15 microns across. However, the true pixel size is 15.24 microns, a difference of one quarter of a pixel, a value that can cause significant errors in precise astrometric measurements. Finally, even more subtle effects such as nonuniformly spaced pixels and improperly produced pixels that differ slightly (∼1%) in size must be considered. The bottom line is, as in all cases where highly precise results are desired, one must know thy CCD in great detail.

Data reduction methods of CCD images from which astrometric measures are to be obtained are similar to those discussed above for general CCD imagery. The differences in the process occur when the final reduced frames are to be used to produce output astrometric information. The interested reader is referred to the discussions and results given in Monet et al. (1991), Lasker et al. (1990a), Monet (1992), and Monet & Dahn (1983).

5.6 Pixel Sampling

An important consideration in photometric and astrometric measurements made with a CCD is how well the PSF is sampled on the two-dimensional array. PSFs that are well sampled by a CCD observation will lead directly to the result that the center and shape of the PSF will be known to higher precision, and thus one will obtain a final answer that will be of higher accuracy. We can define a sampling parameter, r, as follows (Howell et al., 1996; Buonanno & Iannicola, 1989):

$$r = \frac{\text{FWHM}}{p},$$

where FWHM is the full-width half-maximum value of the source PSF and p is the pixel size, both values given in the same units. For r less than about 1.5, digital data are considered undersampled. As can be seen from the above expression, r will be small for the case of a CCD with large pixel sizes compared with the total areal coverage of the imaged PSF. The other possible case of small r values is if the CCD image contains very tight PSFs such as those that might be obtained at observing sites with very good seeing, if using adaptive optics systems, or for CCD images obtained outside the Earth's atmosphere (i.e., space-based telescopes). Real life examples of cases that will produce undersampled images (i.e., small r values) are a typical wide-field telescope outfitted with a large-format CCD, such as a Schmidt telescope or a camera lens, or a space-based telescope such as the *Hubble Space Telescope* wide-field planetary camera (WFPC) (Howell et al. (1996), Holtzman (1990), and Section 7.1).

Anytime the value of r approaches the limiting case of undersampling, standard software methods and techniques of astrometric and photometric data analysis will begin to produce increasingly larger errors and poorer fits as r decreases further (see Section 7.1). Photometric and astrometric errors obtained from CCD observations are related in that the analysis techniques for each type of measurement are very similar. Both photometry and astrometry require intimate knowledge of the centroid position of the source PSF. However, it has been shown (Stone, 1989, Howell & Merline, 1991; King, 1983) that for undersampled data the photometric error is least for source PSFs that are centered on the edge of a pixel or exactly at its center, whereas for astrometric data, the resulting error is least when the source PSF is centered midway between a pixel's edge and its center.

The rule of thumb for pixel sampling on a CCD follows directly from the statistical result of Nyquist sampling. That is, sampling of the PSF of an astronomical source will be optimal in terms of S/N, error rejection, data analysis, and so on for a source PSF that has its FWHM value sampled over about two pixels (i.e., FWHM $\sim 2 \cdot$ pixel size). For example, if the average seeing at an observing site produces source PSFs with FWHM values of 2 arcsec, then an ideal (optimal) CCD pixel size to use would be one for which each pixel within the array has a projected image size of 1 arcsec across.[†] A rigorous mathematical definition of undersampling, based on the Nyquist theorem, identifies critical sampling as the sampling interval that is equal to the width (i.e., standard deviation) of the PSF. For a Gaussian PSF this corresponds to a FHWM equal to 2.355 pixels. Of course an ideal image scale is hard to meet in reality as seeing and telescope focus change with time, source PSFs generally

---

[†] The determination and measurement of CCD pixel size or plate scale was discussed in Section 4.1.

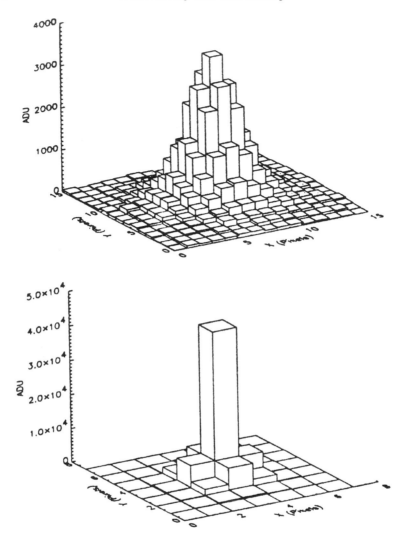

Fig. 5.8. The effects of pixel sampling are shown. The top PSF is a well-sampled star image with a S/N of near 230. The bottom panel shows the same PSF but now severely undersampled and centered at the middle of a pixel and (next page) at the corner of four pixels respectively. Note that the undersampled profiles are not well represented by a Gaussian function. From Howell et al. (1996).

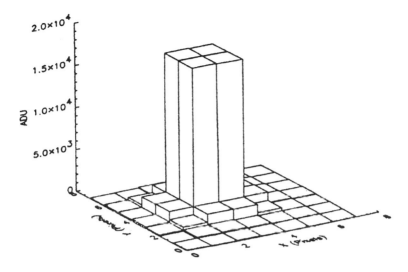

Fig. 5.8. (*Continued*)

do not fall onto the CCD pixel grid exactly on a pixel boundary, and one generally has only a limited number of available CCD cameras with fixed pixel sizes.

Since CCD detectors do indeed sample astronomical sources in a quantized manner, pixel sampling within the array will cause even identical PSFs to change their detailed appearance slightly, even within the same observation. The effects of such sampling differences become worse as the sampling parameter (r) becomes smaller (Merline & Howell, 1995; Howell & Merline, 1991). Figure 5.8 illustrates some examples of various values of r caused by CCD pixel size. The top panel shows a well-sampled source PSF that appears to be more or less Gaussian in shape. The remaining two panels in Figure 5.8 show the same model PSF but now imaged as a poorly sampled ($r = 1$) source. The undersampled cases are for a source PSF with a pixel-centered centroid and a corner-centered centroid respectively. Notice in Figure 5.8 that the undersampled PSFs are not well represented by a Gaussian function (Howell et al., 1996; Buonanno & Iannicola, 1989; Howell & Merline, 1991; Holtzman, 1990).

6

Spectroscopy with CCDs

Although imaging and photometry have been and continue to be main-stays of astronomical observations, spectroscopy is indeed the premier method by which we can learn the physics that occurs within and near the object under study. Photographic plates obtained the first astronomical spectra of bright stars in the late nineteenth century, while the early twentieth century saw the hand-in-hand development of astronomical spectroscopy and atomic physics. Astronomical spectroscopy with photographic plates, or with some method of image enhancement placed in front of a photographic plate, has led to numerous discoveries and formed the basis for modern astrophysics. Astronomical spectra have also had a profound influence on the development of the fields of quantum mechanics and the physics of extreme environments.[†] The low quantum efficiency and nonlinear response of photographic plates placed the ultimate limiting factors on their use.

During the 1970s and early 1980s, astronomy saw the introduction of numerous electronic imaging devices, most of which were applied as detectors for spectroscopic observations. Television-type devices, diode arrays, and various silicon arrays such as Reticons were called into use. They were a step up from plates in a number of respects, one of which was their ability to image not only a spectrum of an object of interest, but simultaneously, the nearby sky background spectrum as well – a feat not always possible with photographic plates. Additionally, and just as important, were the advantages of higher obtainable signal-to-noise ratios, higher quantum efficiency, and very good linearity over a large dynamic range. These advances permitted spectral observations of much fainter

---

[†] Extreme physical environments, such as very high temperatures, magnetic fields, and gravitational fields, are not often available within earthly laboratories.

sources than were previously available. Two-dimensional spectroscopy allowed the large error contribution from the often unknown sky background to be removed during data reduction procedures. However, the above electronic devices still had their problems. Use of high voltage, which caused distortions in the image, and the often low dynamic range both limited the ability to numerically resolve strong spectral lines from a weak continuum or to resolve weak lines in general.

The introduction of CCD detectors for use in astronomical spectroscopy was quick to follow their introduction into the field of astronomy itself. One of the first devices put into general use for astronomical spectroscopy was a Fairchild CCD placed into service at the Coude' feed telescope on Kitt Peak circa 1982. This author remembers that particular device and the remarkable advance it made to astronomy at the time. Observing without photographic plates was amazing.[†] The introduction of CCDs allowed test observations to be made, no chemical development was needed, you could view your data almost immediately after taking it, and mistakes caused you very little in lost time. In addition, fainter sources and better spectral resolution were easily obtained and caused a renaissance in astronomical spectroscopy. The fact that the Fairchild CCD had a read noise of 350 electrons seemed unimportant at the time.

We will begin our discussion of astronomical spectroscopy with point source observations. The term "point source" in generally taken to mean a star, but under various conditions, many objects can be observed as a point source. For example, an active galaxy does indeed often show extended structure in terms of spiral arms, but short exposures or observations intended to study only the nuclear regions are essentially point source measurements. A more formal definition might be that point sources are objects for which their angular size is determined by the seeing disk or instrumental resolution. We will follow point source observations by introducing extended object spectroscopy. The major difference in these two types of spectroscopy is the type of output data product you end up with and the science obtained from the collected data. Our discussion here will concentrate more on the CCD aspects of astronomical spectroscopy and not as much on the actual observational techniques

---

[†] For those of you interested in a bit of nostalgia, remember how one needed to cut the photographic plates completely in the dark, attempt to fit them into the plate holder licking one side along the way to find the emulsion, and then suffering the agony of defeat when you discovered that your 1 hour integration was made with the dark slide closed or the plate had been placed in the holder backwards!

and data reduction procedures. Various types of spectrographs and other related topics are discussed in detail in the excellent reviews given by Walker (1987), Wagner (1992), and Pogge (1992).

6.1 Review of Spectrographs

Spectroscopic observations can be thought of as a method by which one samples the emitted energy distribution from an astronomical source in wavelength bins of size $\Delta\lambda$. Broad-band filter photometry, for example, is a form of spectroscopy; it is merely one with extremely poor spectral resolution. To use spectral information to learn detailed physics for an astronomical object, one must be able to differentiate specific spectral features (lines) from the continuum within the observed spectrum and be able to make quantitative measurements of such features. Generally, this type of analysis requires a spectral resolution of at least 20–40 Å or better. Keep in mind, however, that various scientific objectives can be accomplished with varying amounts of spectral resolution. Schmidt telescope observations using an objective prism and imaging each spectrum onto a CCD have fairly low spectral resolution, but the imaged spectra are indeed useful if the purpose is to identify objects that have blue color excesses (see Section 6.7).

Figure 6.1 illustrates a typical astronomical spectrograph with the common components identified. An entrance slit, onto which the telescope focuses the incoming light from the source of interest, is used to both set the spectral resolution and to eliminate unnecessary background light. An internal light source for the production of a flat field (called a projector flat in spectroscopy) and various wavelength calibration emission line sources are also included. These lamps usually consist of a quartz projector lamp for the flat fielding and a hollow cathode or arc lamp for the calibration sources. Both types of calibration lamp are included in the spectrograph in such a way as to attempt to make their light path through the slit and onto the CCD detector match as closely as possible that of the incoming telescope beam from an astronomical object. Some type of grating (commonly a concave reflection grating) is needed as the dispersive element, although a prism can also be used. Various camera optics, to re-image the slit onto the CCD detector and provide chromatic and field flatness corrections, finish the suite of standard components. Numerous variations on this standard theme have been and will continue to be used as cost, complexity, and purpose of the instrument are always issues.

Fig. 6.1. Schematic diagram of a typical astronomical spectrograph. The major components are the CCD detector, the continuum and comparison line calibration sources, the TV slit viewer, and the grating. From Wagner (1992).

Let us define a few useful quantities in CCD spectroscopy. Table 6.1 lists the various definitions needed for use when discussing the optical properties of the telescope, collimator mirror (or lens), and the camera itself. Using the definitions in Table 6.1, we can define the magnification of the spectrograph as $M = F_{\text{cam}}/F_{\text{col}}$, the projected width of the slit

Table 6.1. *Spectrograph definitions*

D	Diameter of Telescope
F	Focal Length of Telescope
f	Focal Ratio of Telescope
w	Width of Entrance Slit
d_{col}	Diameter of Collimating Mirror
F_{col}	Focal Length of Collimator
f_{col}	Focal Ratio of Collimator
d_{cam}	Diameter of Camera Mirror
F_{cam}	Focal Length of Camera
f_{cam}	Effective Focal Ratio of Camera

at the CCD will be wM, and the slit width will subtend an angle on the sky of $\Theta = w/F$. The projected width of the slit at the CCD detector is then given by

$$r = wM = \Theta FM = \Theta fD\frac{F_{\mathrm{cam}}}{F_{\mathrm{col}}} = \frac{\Theta fDF_{\mathrm{cam}}}{f_{\mathrm{col}}d_{\mathrm{col}}}.$$

To avoid loss of efficiency within the spectrograph, the collimator focal ratio should match that of the telescope. To make this clear, we can write the projected slit width at the CCD detector as $\Re = \Theta Df_{\mathrm{cam}}$. We assume here that the dispersing element does not change the collimated beam size.[†] If the slit is opened wide enough to allow all the light from a point source to pass through, the projected image size of the point source at the CCD detector is simply $\Re$.

The ability to separate closely spaced spectral features is determined by the resolution of the spectrograph. Spectral resolution is defined as $R = \lambda/\Delta\lambda$, in which $\Delta\lambda$ is the difference in wavelength between two closely spaced spectral features, say two spectral lines of equal intensity, each with approximate wavelength λ. Optical light spectral resolutions of a few hundred thousand to one million have been obtained for the Sun, whereas for typical astronomical spectra, R is much less, being near a few thousand down to a few hundred for very faint sources. For comparison, current R values in the infrared are typically less than about 1,000.

For further details of the actual components of astronomical spectrographs, various types of spectroscopy for different applications, and the way in which these components are used to produce spectra at various wavelengths and resolutions, see Wagner (1992), Pogge (1992), DeVeny (1990), Stover et al. (1995), Queloz (1995), Corbally (1995), Cochran (1995), and Robinson (1988b).

6.2 CCD Spectrographs

Current-day spectrographs almost exclusively use CCDs as their detector. The major reasons for this choice are the large free spectral range of modern CCDs (covering roughly 3,000 to 10,000 Å), the linearity of the devices (better than 1% from zero counts to saturation) over a large dynamic range (allowing, for example, detection of absorption or emission lines as well as the continuum), and the large areal format of

---

[†] While not always true for diffraction gratings, this condition is realized for prisms.

modern CCDs (2,048 up to 4,096 pixels or more in extent). This latter property is especially important for applications such as wide-field objective prism work with Schmidt Telescopes, multiobject fiber spectroscopy (Robinson, 1988c) in which the fiber fed spectra are placed in row order on the CCD, and for Echelle spectroscopy for which many spectral orders are two-dimensionally imaged at once (Vogt & Penrod, 1988). The free spectral range obtained could, in principle, be as large as the detector's quantum response, but in practice limits in optical and grating design and CCD size restrict a single spectrograph coverage to somewhere near 4,000 Å or less in the optical bandpass. Some observatories have solved this limited spectral coverage issue by designing and using double spectrographs, with each instrument having two separate spectrograph arms (one red and one blue) with a CCD for each (Oke, 1988). A double spectrograph almost always uses a different type of CCD detector in each arm; each CCD is customized for the best possible detection properties for their particular wavelength coverage. Both high resolution ($R \sim 30{,}000$–$80{,}000$ or more) and low resolution ($R \sim$ a few thousand or less) spectroscopic applications are well suited to using CCDs as the detector; one simply has to match the CCD pixel size to the particular spectrograph and resolution being used.

Optimal sampling of a spectral line that is just unresolved occurs when the FWHM of the line is twice the physical pixel size (Nyquist sampling criteria). It can be assumed in this discussion that a spectral line has an approximately Gaussian shape for which a formal FWHM value can be determined. Note, however, that real spectral lines are not always this well behaved. In addition to matching the spectral line width to the pixel size, CCDs used for astronomical spectroscopy must also have very good charge transfer efficiency (CTE) in order to reduce smearing of spectral lines during readout, which would lead to a loss in spectral resolution. Also, small pixel sizes such as 15 or 9 microns are often desired to meet the Nyquist criteria discussed above.

Let us look at an example for a spectrograph that uses a CCD with 9-micron pixels as the detector. With this setup, the projected slit width size, $\Re$, must be near 18 microns to achieve optimal sampling. For an observing site with typical seeing of 1.5 arcsec, and using a 2- to 5-m telescope, we find (using the formulations given in Section 6.1) that the spectrograph camera must have a focal ratio near unity. This is a very fast focal ratio and requires excellent optical design and near perfect optical surfaces. For the CCD itself, this requirement means that its physical surface must be extremely flat throughout the entire extent of the chip (less than 0.5% rms for accurate spectrophotometry), in order to allow all parts of the spectrum to be in focus simultaneously. As we have seen, this level of flatness can be a difficult requirement to fulfill for certain types of CCD (e.g., thinned devices).

The above example for a CCD spectrograph informs us that, for large-aperture telescopes (say 8–12 m), optimum spectral sampling can only occur if some combination of the following conditions are met. As the telescope diameter increases, the camera focal length must decrease, the seeing disk must decrease, and the detector resolution element (2 CCD pixels) must increase in size. Currently, the hardest requirement to meet in this list is the design and construction of very fast focal ratio cameras. Increasing the CCD pixel size while retaining the large range of total wavelength coverage is the major driving force behind producing even larger format CCDs.

Exceptional seeing, less than 1 arcsec for example, would seem to be the dream of any spectroscopist. However, let us look at an example when very good seeing can cause unexpected results in CCD spectroscopy. The problem is as follows: CCDs are mounted in dewars and attached to the end of a spectrograph in some manner. The dewars are then aligned in an attempt to have the observed spectrum fall onto the detector along either its rows or columns. Perfect alignment of the CCD across the entire spectrum is rarely achieved and thus the imaged spectrum centroid must cross pixel boundarys in the dispersion direction (i.e., as a function of wavelength).

If the object seeing disk becomes less than the projected pixel size, the position of the spectral centroid falls within the pixel itself, alternately occurring at the center of some pixel and then at the pixel boundaries themselves. Wavelength-dependent QE effects within the pixels, due to their gate structures and intrapixel "dead" spots, will cause apparent flux variations that can be as large as $\pm 10\%$ in amplitude. Additional complexities, such as which type of CCD is used, telescope focus, and tracking changes, are harder to quantify and correct for but can have similar effects. The problem described here, that of undersampling, can also occur in CCD imaging applications as well, when the majority of the source PSFs fall within a single pixel. Optimum sampling in CCD imaging also occurs at the Nyquist sampling limit, that is, a point source FWHM should be imaged across about two CCD pixels (see Section 5.6). A several percent error in CCD photometry can occur for images that are undersampled (Howell et al., 1996; Holtzman, 1990).

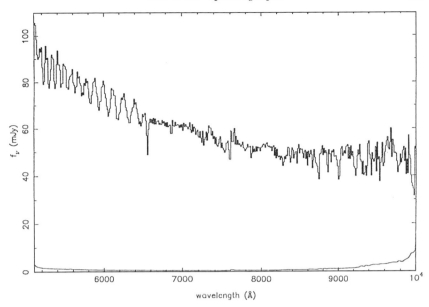

Fig. 6.2. Spectrum of the standard star Feige 92 taken during a time of excellent seeing. This 30-second exposure, obtained with the Faint Object Spectrograph on the Issac Newton telescope, has a pixel scale of 1.2 arcseconds/pixel and this spectrum was obtained when the seeing was sub-arcsecond. The ripple seen in the continuum near 6,000 Å and blueward is due to intrapixel QE differences, which are especially prominent for short wavelength photons in the EEV CCD used. From Dhillon, Rutten, & Jorden (1993).

Figure 6.2 shows an example of the ripple that can occur in a spectrum obtained under conditions of excellent seeing and for which intrapixel QE effects are present (Rutten, Dhillon, & Horne, 1992). Possible methods of correction for spectral ripple include de-focus of the telescope or slightly trailing the spectrum up and down the slit during each integration,[†] neither of which are desirable. Corrections to the spectrum can also be applied after the fact using some empirical model during the reduction phase (Dhillon, Rutten, & Jorden, 1993; Jorden, Deltorn, & Oates, 1993). Spectral ripple, as well as our discussion of pixel sampling in the chapter on photometry, indicates that while very poor sampling is not ideal (as the collected image is spread out over many pixels, each of which adds unwanted noise) undersampling is not necessarily better. Optimum sampling is the best but is not always possible with a given instrument and CCD combination. In addition, the conditions of

[†] This particular solution may remind us long-time observers of a Wobble Plate.

optimum sampling can change with time owing to effects such as seeing, telescope focus, the wavelength of light imaged, and other more subtle effects.

The two-dimensional nature of a CCD allows one to place the spectrum anywhere on the array, targeting good regions of the CCD that avoid manufacturing flaws such as areas of poor sensitivity or dead pixels. Their 2-D design also provides the ability to simultaneously image the (nearby) sky spectrum with the object spectrum. Accurate sky subtraction from a source spectrum increases the S/N of the final result and allows much fainter sources to be observed. It also increases the reliability of the flux measurements and is probably the greatest factor making CCDs superior to other detectors as spectroscopic imagers.

Although large-format CCDs for spectroscopy ($2,048 \times 2,048$, $1,024 \times 3,072$, etc.) are useful in many ways, they have their drawbacks as well. Large CCD formats take longer to read out, make critical areal flat fielding (important for accurate fluxes and line profiles) difficult, and provide more stringent restrictions on mounting and dewar design and overall cost. Physically nonflat CCDs under vacuum (such as thinned devices), shifting of the CCD within the dewar, or movement of the larger, heavier dewar itself can all cause slight movement of spectral features, leading to errors in the analysis. Loss of LN2 throughout the night has been identified as one cause of spectral movement (Smith, 1990a).

Long readout times for large-format CCDs (a few minutes) do not seem like such a big deal compared with typical integrations of 30–60 minutes or more. However, the longer the readout time, the lower the overall observational duty cycle one experiences, and, after obtaining all the necessary calibration spectra, flat fields, comparison arcs, and standard stars, this extra time can become costly. In addition, to produce the final spectrum one often wishes to co-add shorter object exposures to avoid numerous cosmic ray events that can hinder accurate flux and line profile measurements. Obtaining fast temporal coverage with large-format CCDs, such as required for phase-resolved spectroscopy, also gets thwarted, again by these long readout times. In a likewise manner, CCDs that have a limited dynamic range (those with small pixel sixes) force the user to make multiple (shorter) exposures for each type of needed image, especially calibration data for which the high S/N desired can easily saturate the detector.

The finite readout time of large-format CCDs has led to numerous attempts to obtain high-speed spectroscopy via some technique that takes advantage of the two-dimensional nature of the detector. For example, one could step the detector, or the spectrum, along in a direction perpendicular to the dispersion at some predetermined rate, reading out the CCD only after a entire frame of individual spectra is collected. A better idea might be to use the fact that the collected charge can be moved back-and-forth electronically on the CCD from row to row, without actually reading it out. This movement is quite fast, say 200 rows in a few milliseconds, and suffers no readout noise penalty but only a slightly increased dark current. Spectra placed on the top 50 rows of a device will allow faster readout times or one could use the bottom 50 rows with electronic charge movement used to shift the spectrum upward between each integration time. Both of these processes have been tried with success (Robinson, 1988b).

One could even slosh the charge back-and-forth periodically in order to build up the signal in two spectra for say a given phase of a binary star. The spectrum not being exposed gets hidden under a mask and the final image produces two or more "phased" spectra. An idea such as this was developed for spectropolarimetry (Miller, Robinson, & Goodrich, 1988) with some success and led to the discovery of some previously unknown, yet interesting CCD effects (Janesick et al., 1987a; Blouke et al., 1988).

6.3 CCD Spectroscopy

Two common measurements from an astronomical spectrum are those of the flux level as a function of wavelength and the shape and strength of spectral lines. Currently, the best determined absolute fluxes are near 1%, usually for spectral data covering a small wavelength range and for bright (high S/N) sources (e.g., Tüg, White, & Lockwood (1977), Furenlid & Meylon (1990), Hayes & Latham (1975), Oke & Gunn (1983), and Schectman & Hiltner (1976)). Absolute flux observations must make use of a wide slit or hole (10 arcsec or larger in size is typical) as the entrance aperture to the spectrograph. Use of such a wide slit assures that 100% of the source light is collected and avoids problems related to telescope tracking or guiding errors and differential refraction (see below).

In principle, relative flux measurements should be determinable to even better levels than those quoted above and the shape of a spectral line should be almost completely set by the instrumental profile of the spectrograph. Spectroscopy desiring relative fluxes generally makes use of a narrow slit (matched to the seeing) to preserve the best possible spectral resolution. For example, for an observation requiring accurate radial velocity measurements a narrow slit would be preferred to achieve the highest possible spectral resolution. After all, the imaged spectral features (lines) are merely images of the slit itself focused onto the CCD.

Calibration of observed object spectra is normally performed by obtaining spectra of flux standard stars with the same instrumental setup including the slit width. During the reduction process, these standard stars are used to "flux" the object data, that is, assign relative fluxes to your object spectrum counts as a function of wavelength. It is assumed that during each integration, the object of interest and each standard star sent the same relative color and percentage of their PSF light per time interval through the slit and onto the CCD. Seeing changes, nonphotometric conditions, guiding errors, and differential refraction can all negate this assumption.

Three factors unrelated to the CCD detector itself – tracking and guiding errors, seeing changes, and spectrograph slit angle – are important to your final spectral result in the following ways. Observations of an astronomical source made through a spectrograph slit are often obtained such that the slit size matches the object seeing disk. That is, the slit width is set to allow most of the PSF (for a point source) to pass through but kept small enough to eliminate as much of the sky background as possible from entering the spectrograph. Therefore, changes in telescope guiding or image seeing will cause slightly more or less source light to enter through the slit. These effects can cause large noticeable effects as well as more subtle possible unknown effects to be part of your final image. These issues, as well as items such as the use of a red sensitive slit viewing/guiding camera when observing very blue sources, will not be discussed further here (Wagner, 1992).

The angle on the sky of the spectrograph slit is very relevant to the use of CCDs as detectors. Observations of an astronomical source at some angle away from the zenith (an airmass of 1.0) can cause differential refraction to become an issue. Differential refraction is the variation of the angle of refraction of a light ray with zenith distance. All objects in the sky become slightly prismatic owing to differential refraction by the Earth's atmosphere. If the spectrograph slit is not placed parallel to the direction of atmospheric dispersion, nonchromatically uniform light loss may occur at the slit since the image will have an extended, color-dependent PSF. Atmospheric dispersion is caused by the variation of the angle of refraction of a light ray as a function of its wavelength, and the direction parallel to this dispersion is called the parallactic angle. Using spherical trigonometry, the parallactic angle can be determined from the following:

cos(object declination) × sin(parallactic angle)

= sign(hour angle) × cos(observers latitude) × sin(object azimuth),

where sign = +1, if the hour angle is positive and −1 if it is negative.

Table 6.2 presents an example of the amount of differential refraction that a point source will experience, as a function of airmass, for an observatory at an elevation of 2,200 m. Note that all values in the table are relative to 5,500 Å and are of significant magnitude at essentially all wavelengths and airmasses. CCD observations are particularly sensitive to an effect such as differential refraction as their very good red or blue QE could produce spectral data with highly inaccurate flux values. Even more troublesome is the case of using, for example, a blue flux standard star when observing red sources. Differential refraction can cause the light entering the slit, and therefore imaged by the CCD itself, to record incorrect intensities as a function of wavelength. Aligning the slit at the parallactic angle solves this problem.

A fun example to consider concerning differential refraction is the possible use of a direct CCD imager at very high airmass. The imager will act like a spectrograph and record spectra of sources using the Earth's atmosphere as the dispersive element. Some simple calculations will reveal that while possibly a good idea in principle, at the very high airmasses needed, the dispersion is extremely nonlinear and changes very rapidly with time. Extremely short exposures would be required, thereby allowing data to be gathered for only the brightest of stars. In addition, most astronomical telescopes can not point to the needed positions to obtain these sorts of very high airmass observations.

Some additional considerations related to the use of CCDs in astronomical spectrographs are those of order sorting filters, stray light, fringing, and economics. The first three of these cause undesired light to be imaged onto the CCD, some of which falls directly on top of the object spectrum of interest. This extra light increases the apparent background, decreases the resulting S/N, and is often difficult or impossible to remove or even measure. As an example, using a diffraction grating to observe blue light in second order necessitates removal of the first-order red spectral light from the incident beam before detection by the CCD. Since CCDs are generally very red sensitive, this is a critical step to perform.

Table 6.2. *Example Differential Refraction Values (in arcsec)*

Altitude	Airmass	3000 Å	3500 Å	4000 Å	4500 Å	5000 Å	5500 Å	6000 Å	6500 Å	7000 Å	8000 Å	9000 Å	10000 Å
90	1.00	0.0	0.0	0.0	0.0	0.0	0.0	0.0	0.0	0.0	0.0	0.0	0.0
75	1.04	0.6	0.35	0.2	0.1	0.1	0.0	0.0	−0.1	−0.1	−0.1	−0.1	−0.2
60	1.15	1.3	0.8	0.5	0.2	0.1	0.0	−0.1	−0.1	−0.2	−0.3	−0.3	−0.4
45	1.41	2.3	1.45	0.8	0.4	0.2	0.0	−0.1	−0.2	−0.3	−0.5	−0.6	−0.6
30	1.99	4.0	2.45	1.4	0.7	0.3	0.0	−0.2	−0.4	−0.6	−0.8	−1.0	−1.1
20	2.90	6.3	3.9	2.3	1.2	0.5	0.0	−0.4	−0.6	−1.0	−1.3	−1.5	−1.7
15	3.82	8.5	5.3	3.1	1.6	0.7	0.0	−0.5	−0.9	−1.3	−1.7	−2.1	−2.3
10	5.60	12.9	8.0	4.7	2.4	1.1	0.0	−0.8	−1.3	−2.0	−2.6	−3.1	−3.5
5	10.21	26.1	16.2	9.5	4.9	2.2	0.0	−1.6	−2.7	−4.0	−5.2	−6.3	−7.0

A CuSO$_4$ filter is probably the best order sorting filter to use in this circumstance but is far from ideal. CuSO$_4$ filters have poor UV and blue transmission, are inconvenient to use (they are either a crystal or a liquid filter), and have a long wavelength red leak. Further discussion of such intricacies can be found in books concerning spectrographic observations of astronomical sources; also see Wagner (1992), Pogge (1992), and Robinson (1988b).

Finally, we conclude this section with our wish list for the ideal CCD to use within an astronomical spectrograph. The CCD should have low read and dark noise, high QE over a large wavelength range, very good CTE, small pixel size, and large dynamic range, and it should be "tuned" to the application desired. Tuning a CCD simply means that one should use the device that is best suited for the job at hand. Properties such as high red sensitivity, very deep pixel wells, small pixel size, back-side illumination and coating, etc. are items worthy of consideration.

6.4 Signal-to-Noise Calculations for Spectroscopy

Calculation of the signal-to-noise ratio for spectroscopic observations is performed in a manner similar to that described earlier in this book (Chapter 4). As in photometric measures, we find that for bright sources $S/N \propto \sqrt{N}$ while for faint sources we must use the entire S/N expression (see Section 4.4). For spectroscopic observations obtained with a good CCD system, the largest noise contributors that will degrade the resulting S/N are the background sky contamination and how well the data can be flat fielded. The value of the S/N of a spectroscopic observation can have a few different meanings.[†] For CCD spectroscopy, one can calculate the S/N for the continuum or the S/N for a given spectral line.

In the continuum case, the number of pixels, n_{pix}, used in the S/N calculation will be determined by the continuum bandpass range over which the S/N is desired times the finite width of the spectrum on the CCD. For example, a typical CCD spectrograph might have an image scale of 0.85 Å/pixel and the imaged spectrum may have a width of 3 pixels on the array in the direction perpendicular to the dispersion. To calculate the S/N for the spectral continuum over a 100 Å bandpass

---

[†] Note here that this is also true for CCD imaging or photometric applications as well. Depending on the user's choice of parameters such as n_{pix}, the final S/N value will change. Likewise, reported S/N values without comment on the exact choice of specific parameters are, at times, difficult to interpret.

in this example, one would use the value of 353 for $n_{\rm pix}$. In contrast, a narrow emission line with a full width at zero intensity (FWZI) of 40 Å would use an $n_{\rm pix}$ of 141. The S/N of the emission line will therefore be higher in value owing to the smaller overall error contribution (approximately 3 to 1) not to mention the higher flux values per pixel within the line itself. As an exercise, the reader might consider how one would calculate the S/N in the case of an absorption line.

Signal-to-noise calculations are also useful in predicting observational values such as the integration time needed to obtain a desired scientific result from your spectroscopic observations. This type of calculation can be performed using the formulae presented in Section 4.4. Spectroscopic S/N values in the continuum of near 10 are often sufficient for gross spectral classification, whereas values in excess of 500 are needed for detailed abundance analysis or weak absorption line measurements. When making predictions of the S/N to expect in a spectroscopic observation, keep in mind that spectrographs are much less efficient overall than direct imaging cameras (2–4% vs. 30–40%) and that seeing effects and slit losses can be considerable in terms of the actual flux collected by the spectrograph.

6.5 Data Reduction for CCD Spectroscopy

This section discusses the basics of astronomical CCD spectroscopic data reduction. Figure 6.3 presents a raw, unprocessed CCD spectroscopic image of a point source. The initial reduction steps for CCD spectroscopy are exactly the same as previously discussed for imaging applications. Bias (or dark) frames and flat field calibration images are needed and used in the same way. After performing these basic reduction procedures for the CCD images, there are additional steps one must take that are specifically related to spectroscopy. These extra steps involve the use of spectra of spectrophotometric flux standards and wavelength calibration (arc) lamps. Figure 6.4 illustrates the five types of CCD images needed for complete reduction and calibration of spectroscopic observations. We will not discuss some minor, yet important processing steps such as cosmic ray removal, bad pixel replacement, and night sky emission line complexities. Detailed instructions for CCD spectroscopic data reduction can be found in Pogge (1992) and Massey, Valdes, & Barnes, (1992). The article by Wagner (1992) is particularly useful, being the best review of the subject to date.

Fig. 6.3. Raw CCD image of a point source. The spectrum runs vertically up and down near the middle of the CCD with wavelength increasing toward the top. Night sky emission lines are clearly seen running horizontally across the entire frame. The sky lines due to OI at 5,577, 6,300, and 6,363 Å are visible at the bottom, and numerous OH airglow lines are seen near the top of the image. The large-scale gradients are due to nonuniform CCD illumination within the spectrograph. From Wagner (1992).

For a point source, the final product desired is a 1-D spectrum consisting of wavelength versus relative or absolute flux. It is assumed that you have in hand CCD spectra of your object(s) of interest, at least one observation of a flux standard, and wavelength calibration spectra of arc lamps (see Figure 6.4). The first step with the reduced two-dimensional CCD image, after the standard processing with bias and flats is performed, is to extract the spectrum itself and collapse it into

Fig. 6.4. The five necessary types of CCD image needed for spectral reduction. From bottom to top we see a bias frame, a flat field frame, the object spectrum of interest, the spectrum of a flux standard, and finally a FeNe arc spectrum. The flux standard will be used to turn observed counts in the object into relative flux, and the arc spectrum will be used to set the wavelength scale. From Wagner (1992).

a 1-D image of pixel number versus ADU/pixel. In the simplest (and unrealistic) case, the imaged spectrum lies exactly along one row (or column) of the CCD and one can extract it simply by extracting the single row from the final image. Generally, the imaged spectrum covers two or three or more rows (or columns) and the extraction process involves some manner of summing a few adjacent rows perpendicular to the dispersion (an "after-the-fact" pixel binning) at each point along the dispersion. Furthermore, in practice it is often found that CCD spectra are not precisely aligned with the CCD pixels and are curved on the detector as the result of the camera optics, instrumental distortions, or CCD flatness issues. Fainter sources present additional complexities, as centroiding the spectrum (for, by example, using cuts across the dispersion direction) in order to extract it is often difficult or impossible. A typical example might be a spectral observation of a faint continuum source that contains bright emission lines. Details of the spectral extraction process sky subtraction and optimal extraction techniques are discussed in Wagner (1992), Pogge (1992), Robinson (1988b), Robinson (1988c), Schectman & Hiltner (1976), and Horne (1986).

At this point, your extracted spectrum will have an x axis that is in pixels and we wish to convert this to wavelength. The procedure to perform such a task involves observations of calibration arc spectra obtained often during your observing run. The idea is to match the x-axis pixel scale of the calibration arc lamps with their known wavelengths and then apply this scaling procedure to your object data. Calibration arc spectra of sources such as hollow cathode Fe lamps or He-Ne-Ar lamps contain numerous narrow emission lines of known wavelength. Use of these emission lines during the reduction procedures allows a conversion from a pixel scale to a (possibly rebinned linear) wavelength scale. Correction for atmospheric extinction (Hendon & Kaitchuck, 1982), similar in manner to photometric corrections already discussed but generally a bit more complex owing to the larger wavelength coverage, can now be applied to all obtained spectra. Since you are relying in this step on the hope that the arc emission lines fall onto specific CCD pixels and define a wavelength scale that is identical to that of your object, you want to obtain arc data in as similar a manner to that used to gather your object spectra as possible. Instrument flexure caused by telescope motion throughout the night or movement of the CCD within the dewar are two of many possible effects that will invalidate the wavelength to pixel scaling procedure.

We now have to deal with our collected spectra of flux standards. These stars are observed solely for the purpose of converting the collected pixel count values into absolute or relative flux values. Application of all of the above steps to your spectra of spectrophotometric flux standards will produce 1-D data with a wavelength scale (x axis) versus counts on the y axis. We now wish to have the y axis of ADUs or counts converted into flux units such as ergs s^{-1} cm^{-2} Å^{-1}. Most observatories and data reduction packages (such as IRAF and MIDAS) contain lists of spectrophotometric flux standards appropriate to observe and use for fluxing of your spectroscopic data. Within the reduction software, tables of wavelength versus flux are kept for a large number of spectrophotometric flux standards. To understand and appreciate the details involved in setting up even a single spectrophotometric standard star, see Tüg, White, & Lockwood (1977), Jacoby, Hunter, & Christian (1984), and Massey et al. (1988). In a similar manner to the method by which we took the known arc wavelengths and converted their pixel scale into a wavelength scale, we can now take the known fluxes of the standard stars and convert pixel counts into relative or absolute fluxes. The difference between relative and absolute is essentially the difference between a narrow or large slit width as mentioned above. The conversion of counts to flux is performed under the assumption that slit losses, color terms, transparency, and seeing were similar between the standard star observations and the object(s) of interest.

One can never have too many calibration data and must always trade off time spent collecting CCD frames of standard stars and arcs with collection of data for the objects of interest. Instrument flexure, non-photometric conditions, color terms, and accurate wavelength calibration are crucial to the production of accurate final spectroscopic results such as that shown in Figure 6.5.

6.6 Extended Object Spectroscopy

The definition of extended object spectroscopy follows from the fact that we wish to obtain spectra, not from an unresolved point source, but from a region of the sky for which we desire simultaneous wavelength and spatial information. Examples might include galaxies, nebulae, and planets within our solar system. While there is no fundamental difference between this type of spectroscopy and point source observations such as those described above, their are differences in the instruments used and the reduction techniques involved. We will present here a very basic introduction to the subject and refer the reader to the more detailed review given by Pogge (1992).

One method of obtaining spectra of an extended object is by using long-slit spectroscopy. While sounding like an entirely new method of observing, long-slit spectroscopy is very basic. When we discussed

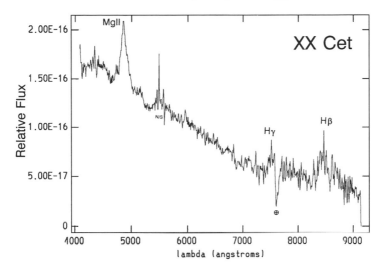

Fig. 6.5. Final reduced CCD spectrum of the quasar XX Cet. This object, long suspected to be a cataclysmic variable, shows redshifted spectral lines (redshift = 0.736), the telluric A band due to the Earth's atmosphere, and false very narrow spectral lines (N/S) due to incomplete night sky emission line subtraction during the reduction process. The flux is in units of ergs s^{-1} cm^{-2} Å^{-1}. From Howell & Usher (1993).

point source observations, we concerned ourselves with the details of the spectrograph and final spectrum as they related to the incident light from a point source focused on the spectrograph slit and imaged on the CCD array. We mentioned in that discussion that we desired to use the slit to keep out as much background light as possible. However, all objects that lie along the slit, say in the x direction as presented in Figure 6.6, will produce a spectrum and be imaged on the 2-D CCD array. This last statement makes some assumptions about the ability of the spectrograph optics and optical path to support such an endeavor.

One can imagine a case in which a number of point sources are lined up in an exact east–west manner and a spectroscopic observation is performed of these stars using an east–west slit alignment. The output image will contain a number of parallel spectra, one for each point source, lined up on the CCD. Obvious extensions of this simple example to real cases include placing the long slit of a CCD spectrograph along the axis of a spiral galaxy or alternately across and alongside of bright knots of ejecta within a recently erupted classical novae. The uses of such

Fig. 6.6. Schematic view of a typical long-slit CCD spectrograph. Positions along the slit are mapped in a one-to-one manner onto the CCD detector. A number of optical elements in the camera, used to re-image and focus the spectrum, have been omitted from this drawing. From Pogge (1992).

spectrographic techniques are many and yield information on both the spectral properties of an object (radial velocity, line widths, etc.) and the spatial properties (spiral arms versus the bulge, etc.). This type of observational program is indeed another that benefits from large-format CCDs.

A more complex and upcoming type of two-dimensional CCD spectrograph is the imaging Fabry–Perot interferometer. Within this type of a spectrograph, 2-D spectral imaging over a narrow bandpass can be collected on the CCD, and with a change in the Fabry–Perot etalon spacing, a new narrow bandpass can be isolated. These types of instruments are often quite complex to master and setup, produce complex data cubes of resultant imagery, and tend to be less efficient than conventional spectrographs owing to their many optical surfaces. However, to its benefit, Fabry–Perot interferometry has a unique capability to provide tremendous amounts of scientific data per observational data set and to obtain fantastic results for the objects observed. The details of this type of spectrograph and its use can be found in Pogge (1992), Bland & Tully (1989), and Roesler et al. (1982).

A final method of obtaining spatial as well as spectral information is that of narrow-band imaging. The spatial aspect comes about through the use of a CCD as a wide-field imager and the spectroscopic information is provided at selected wavelengths by the use of narrow-band filters. This is no different than observing a given area of the sky in B, V, and R filters, except that here one usually is imaging one or more extended objects and selects carefully two or more restrictive filters in order to obtain specific results. Simultaneous imaging in all colors is lost as one has only limited spectral coverage with this technique, but one obtains full spatial sampling over the entire the field of view and the instrumental setup and data reduction are quite simple. In addition, this type of 2-D spectrophotometry can be performed with any CCD imager with the addition of the appropriate filters.

There are numerous other complexities associated with extended object spectroscopy, some in the observational setup procedures needed and some in the data handling and reduction procedures. However, the wealth of information available from such observations far outweigh the difficulties of these more complex types of CCD spectroscopy.

6.7 Slitless Spectroscopy

The above discussion of astronomical spectroscopy was based on the use of a spectrograph that allowed an image of a slit to be dispersed and then re-imaged onto the CCD detector. Another method, slitless spectroscopy, has also been a workhorse for astrophysical science (Bowen, 1960a) and continues to be useful today (MacConnell, 1995). The idea of slitless spectroscopy is to take advantage of a simple direct imaging system to provide a mechanism for producing and recording astronomical spectra. Some form of dispersive element is placed either before the telescope entrance aperture (e.g., an objective prism) or just before the detector (e.g., a grism), allowing a spectrum to be formed and imaged on the detector.

Work with objective prisms placed in the optical path of a Schmidt telescope and the resulting slitless spectra being imaged on large-format photographic plates has been used in astronomy for over seventy years. This type of an arrangement allows one to obtain spectra of all imaged objects in the field of view with the integration time setting the limit for the faintest usable spectra (Walker, 1987). The resulting spectra are generally of low dispersion and used for gross classification purposes (MacConnell, 1995), but even radial velocities can be measured (Fehrenbach, 1967). The dispersions available with an objective prism arrangement are near 200–300 Å/mm and the spectra obtained generally cover a total bandpass of 400–1000 Å. The limiting magnitude for a usable spectrum with an objective prism setup is approximately given by

$$m_{\text{lim}} = 18.5 + 5\log_{10}T,$$

where T is the focal length of the telescope in meters. Generally, this limit is 8–10 magnitudes brighter than that available at the same integration time with the same optical/detector system when used as a direct imager.

Objective prisms have the disadvantage of being massive and producing nonlinear dispersions. Transmission gratings have replaced objective prisms in some applications although they can introduce coma into the final image. To counter the effect of coma, grating-prism combinations or grisms were developed (Bowen, 1960b). A variant of this idea is the grens or a grating-lens combination. Gratings can be blazed to produce high throughputs in a single order, but the zeroth-order image is always present and leads to field crowding, background light issues, and possible confusion as an emission line in another spectrum (Schmidt, Schneider, & Gunn, 1986).

Current work with slitless spectroscopy of interest here are those applications that use a CCD as a detector. Such examples are the PFUEI (Schmidt, Schneider, & Gunn, 1986; Gunn & Westphal, 1981) and COSMIC (Kells et al., 1998) instruments used at the 200″ Hale telescope and the QUEST instrument (Sabby, Coppi, & Oemler, 1998) used on a 1-m Schmidt telescope. These instruments use a transmission grating, a grism (preceded by a multislit aperture mask), and an objective prism respectively to form their spectra, which are then imaged onto a CCD. The above formula for the limiting magnitude of a useful spectrum should be revised downward by 1–2 magnitudes for a high throughput, high QE, low noise CCD slitless spectroscopy system.

All types of slitless spectroscopy must cope with three similar issues in terms of their image acquisition and data reduction. First, it is often useful to obtain a direct image of the field without the dispersing element in place. This image allows accurate source location independent of intervening spectral overlap or zero-order confusion (in the case of using a grating). Slitless spectra can cover many hundreds of pixels on the CCD with the zeroth-order image being separated by 100–200 pixels. Second, data extraction from the final image can often be complex as field crowding or seeing changes during an exposure will cause broadened, ill-defined spectra. Procedures here generally use a rectangular extraction box containing the spectrum, yet being small enough to avoid large sky contributions. Finally, there is the issue of calibration of the spectrum in flux and wavelength. Wavelength calibration is usually accomplished by one of two methods: Observation of an astronomical source with well-defined, known spectral (usually emission) lines or the use of a calibration lamp shone through pinholes placed at the focal plane. Other methods include use of known spectral lines in one or more of the imaged spectra or the centroid of the zeroth-order image[†] and a knowledge of the plate scale to calculate the dispersion per CCD pixel (Schmidt, Schneider, & Gunn, 1986).

Of the above three concerns, the calibration of the obtained spectra is the most important and the hardest to perform. It is often the case that slitless spectra are presented as unfluxed, having a spectral shape dominated by the instrument and detector response, and often with a wavelength scale that is only approximate. The latter is usually sufficient, as dispersions of 10–50 Å/pixel or more provide wavelength resolutions of only 10 to 200 Å or so. Thus, precise wavelengths are often unimportant. Providing fluxes for the spectra is a difficult and often ignored issue (see below), although not always (Schmidt, Schneider, & Gunn, 1986; Schmidt & Gunn, 1986). Spectra classification, identification of (redshifted) emission lines, or separation into blue and red objects are typical goals, none of which require more than relative flux values.

Two considerations in the determination of the flux from an object imaged via slitless spectroscopy are background subtraction and image flat fielding. The determination of the sky background in a CCD image

---

[†] The zeroth-order image, while possibly a nuisance, can be used to obtain photometric information such as an instrumental magnitude or, for time-series slitless spectroscopy, a light curve.

of slitless spectra can be confusing. While all the light from the astronomical objects is passed through some sort of dispersing element, so is the light from the background sky. Thus, at any point on the CCD detector, the background is a combination of spectrally dispersed but unresolved sky light. Thus, a "sky" section of the CCD lying adjacent to an object spectrum is not a true representation of the (dispersed) background level imaged with the spectrum. Sky subtraction is not a simple procedure and is often not performed.

Flat fielding a slitless image also presents challenges. We discussed above how the color terms in a flat field image can have large effects on the outcome of the calibration procedure. You can imagine that color terms are even more important in this circumstance, as the flat field light is imaged and dispersed across the CCD in a hopefully similar manner to that of your data. As expected, sky flats provide a far better flat field than dome flats (Schmidt, Schneider, & Gunn, 1986).

Let us examine a few slitless spectroscopy systems in detail. The prime focus universal extragalactic instrument (PFUEI) (Schmidt, Schneider, & Gunn, 1986; Gunn & Westphal, 1981; Schmidt & Gunn, 1986) uses a TI CCD with 15-micron pixels, an image or plate scale of 0.4 arcsec/pixel, and a field of view of 30 arcmin. A transmission grating with 75 lines per mm, blazed at 6,500 Å, is used to provide a spectral dispersion of 35 Å/pixel. The transmission grating can be easily removed from the optical path allowing a direct CCD image to be obtained. When placed in the beam, spectra of each object within the field of view are recorded on the CCD. Figure 6.7 shows a portion of a typical slitless spectroscopic CCD image obtained with the PFUEI.

An innovative setup can combine CCD drift scanning techniques with slitless spectroscopy. The QUEST instrument [Sabby, Coppi, & Oemler (1998) and Section 4.6] is such a camera operating on a 1-m Schmidt telescope and using a CCD array of 16 2,048 × 2,048 Loral CCDs. A 3.2 degree objective prism is employed to provide 401 Å/mm dispersion, yielding a spectral resolution of ∼10 Å. A cutoff filter is used to stop light longward of 7,000 Å. This cutoff filter reduces the increased red-ward sky background as well as shortening the length of the imaged spectra, thus decreasing the chance of spectral overlap. Since a prism is used as the dispersing element, no zero-order image concerns exist; however, the spectra do have a nonlinear dispersion.

Figure 6.8 shows a portion of a QUEST drift scan with the prism in place. The spectra are about 400 pixels long and have the dispersion direction aligned with the scan direction to avoid spectral clipping at

Fig. 6.7. A portion of a PFUEI spectroscopic CCD image. The figure is 5.5 arcmin on a side with east at the top and north to the left. Note the zeroth-order (pointlike) images for each spectrum that appear about one spectral length to their west. The large starlike object near the top middle of the frame is a saturated zeroth-order image from an object whose spectrum is off the top of the image. The zeroth-order images are at the positions of the sources as seen in a direct image and are used in the PFUEI data reduction to set the wavelength scale. The arrow points to the C IV 1550 Å line in a redshifted 19.5 magnitude quasar. From Schmidt & Gunn (1986).

the CCD edges. No attempt is made to flux calibrate the data but initial wavelength calibration is accomplished by identification of known spectral lines detectable in stars.

Slitless spectroscopy is a wonderful tool for obtaining spectra of many objects simultaneously and to very faint limits. The simplistic nature

Spectroscopy with CCDs

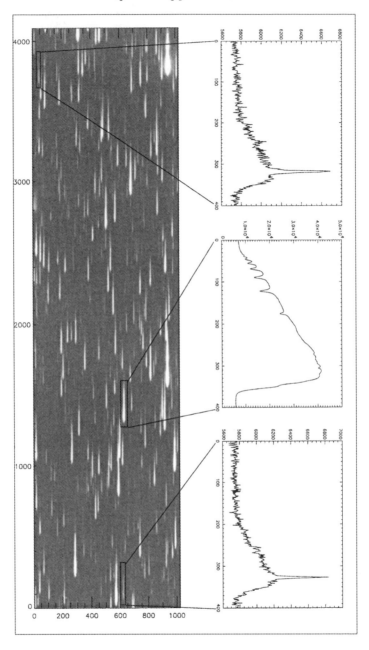

Fig. 6.8. A 1,000 × 4,000 pixel section of a QUEST CCD drift scan obtained with the use of an objective prism. The dispersion is parallel to the scan direction and each spectrum is 200–400 pixels in extent. Examples of some extracted spectra are shown on the right. From Sabby, Coppi, & Oemler (1998).

of the technique makes for an ease of use given that a direct imaging CCD camera is available. The addition of slitless spectroscopy with large-format CCDs and CCD drift scanning go even further, allowing spectroscopic identification of many thousands of objects with little additional effort beyond the direct imaging already in place. Slitless spectroscopy promises to be an evolving CCD application, yet to have reached its peak potential.

7

CCDs Used in Space
and at Short Wavelengths

The current high level of understanding of CCDs in terms of their manufacture, inherent characteristics, instrumental capabilities, and data analysis techniques make these devices desirable for use in spacecraft and satellite observatories and at wavelengths other than the optical. Silicon provides at least some response to photons over the large wavelength range from about 1 to 10,000 Å. Figure 7.1 shows this response by presenting the absorption depth of silicon over an expanded wavelength range. Unless aided in some manner, the intrinsic properties of silicon over the UV and EUV spectral range (1,000–3,000 Å) are such that the QE of the device at these wavelengths is typically only a few percent or less. This low QE value is due to the fact that for these very short wavelengths, the absorption depth of silicon is near 30–50 Å, far less than the wavelength of the incident light itself. Thus, the majority of the light (~70%) is reflected with the remaining percentage passing directly through the CCD unhindered.

Observations at wavelengths shorter than about 3,000 Å involve additional complexities not encountered with ground-based optical observations. Access to these short wavelengths can only be obtained via space-based telescopes or high altitude rocket and balloon flights. The latter are of short duration from only a few hours up to possibly hundreds of days and use newly developing high-altitude ultra-long duration balloon flight technologies. Space-based observations in the high energy regime from UV to shorter wavelengths usually require detectors to be "solar blind." The term solar blind means that the detector must be completely insensitive to visible light photons. This is generally accomplished by using a nonoptically active type of detector or through the use of various types of filters. The majority of astronomical objects emit 10^4–10^6 visible light photons for every UV or shorter wavelength photon; thus even

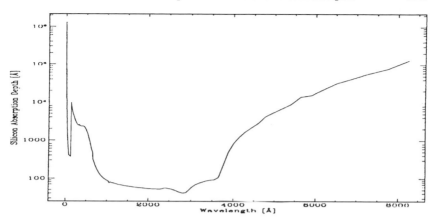

Fig. 7.1. Silicon absorption depth (in Å) from 1.2 to 8,500 Å. The vertical axis is a log scale with major tick marks starting at 100 and ending at 10^6. From Bonanna (1995).

a visible light blocking filter with a 1% UV transmission is not nearly sufficient to remove optical contamination. In addition, most common filters used to block visible light also absorb some of the incident higher energy radiation as well. Use of such absorbing filters cause even a high QE CCD at UV wavelengths (say 20%) to be reduced to a low effective QE near 2%.

Long-term exposure to high vacuum can cause contamination of the dewar in which the CCD resides. This contamination can be through outgassing of various materials such as vacuum grease or AR coatings, normally not a problem in well-produced ground-based systems. Exposure to high energy radiation can cause changes in the QE of the CCD or cause permanent damage to the pixels and electronic structures within the array. Studies of the effects of high energy radiation and space environment observations on CCDs are ongoing at a number of laboratories, the most famous probably being the Jet Propulsion Laboratory (Janesick & Elliott, 1992; Janesick, Hynecek, & Blouke, 1981; Janesick, Elliott, & Pool, 1988; Holtzman, 1990).

Before we discuss the details of observations at wavelengths shorter than the optical, we need to make a brief detour to look into some special issues related to space-based observations with CCDs. The more notable of these issues are the calibration of the CCD throughout the instrument or mission lifetime, the fact that the point spread function is much smaller than generally obtained with ground-based data, and continual degradation of the CCD with time as the result of radiation damage.

7.1 CCDs in Space

Space-based CCDs have a number of special problems associated with them that are often not considered for ground-based systems. Once launched, human intervention is unlikely and the CCD and instrument package can never be retrieved for fault correction or calibration purposes. Even simple procedures, such as bias calibration, take on new meaning as CCD evolution or changes in the gain or other CCD electronics mean new calibration images are needed. Damage to the array (see Section 7.2), or the possibility that the primary circuits fail and the backup electronics or even a different clocking scheme must be used, means that new calibration images must be produced. Also, each observer does not have the ability to obtain all the needed calibration data and the project must provide the finest and most up-to-date calibration images for each CCD, instrument and mode of operation. All issues have to be thought out completely prior to launch or dealt with through analysis of downloaded data during the mission.

One such example of a significant change in CCD operation is provided by the *Hubble* WFPC2 instrument (Holtzman et al., 1995b). After operating in space for only about three months, it was noticed that the CCDs developed an odd sort of CTE effect. The effect caused stars to appear fainter if imaged in higher numbered rows. The apparent cause was the development of a large number of traps within the CCDs not seen during preflight tests. Photometric gradients of 10–15% were present along CCD columns and, even worse, the effect was highly dependent on the brightness of the imaged star, being only about 5% for bright stars.

Using ground-based laboratory tests with similar electronics and CCDs, it was determined that changing the operating temperature from $-76°$C to $-88°$C would cause a sharp decrease in the CTE effect. Such a change caused the CTE variations to almost disappear, leaving only a 3–4% gradient. A further temperature decrease would probably improve the situation but in-flight hardware did not allow the CCDs to be operated at colder levels. Thus, considerable effort has been put into the development of a semi-empirical software model that can be applied to data obtained with the WFPC2 in order to correct for the remaining effect (Holtzman et al., 1995a; Whitmore & Heyer, 1998).

As a consequence of the CCD operating temperature being lowered, a decreased dark current was observed. However, on-orbit hot pixel development was greater than expected with many of these hot pixels "fixing" themselves after dewar warming (Section 7.2). Calibration dark frames are therefore required often to monitor the dark current and to provide the best dark frames to use given any set of observational circumstances. Hot pixels are especially important to understand in space-based CCD imagery as the very small PSF of imaged scenes and the appearance of numerous cosmic rays with a plethora of shapes, including single pixel events, must be distinguished from the collected flux of interest.

We alluded above to the importance of cosmic ray identification in order to avoid misinterpretation of imaged scenes. For a sample of 2,000 second dark images taken with the WFPC2 it was found that 5–10% of the cosmic ray events were single pixel events of 5 sigma or greater above the bias level. Fully one half or more of these events showed consistent pixel positions from frame to frame and thus could not be identified with true cosmic rays or local radioactivity from the dewar and surroundings. Typical signal levels for true single pixel cosmic ray events were near 200 electrons while multiple events peaked near 700 electrons (Holtzman et al., 1995b). Multiple pixel cosmic ray hits (averaging 7 affected pixels per event) are much more common than single pixel events, and a rate of almost 2 events per CCD per second was seen.

CCD dewars, once sealed, evacuated, and chilled, are often seen to produce contaminants owing to outgassing of grease or other coatings used in their construction. When at operating temperatures of −80°C or so, the dewar window is a good site for condensation of such contaminants. These small particles of material are very good absorbers of light, particularly UV radiation, because of their characteristic sizes.

One simple calibration test that allows monitoring of this effect is to obtain fairly regular observations of a bright UV star. If the dewar window does indeed get fogged with material, careful measurements of the UV throughput of the observed flux will show a slow degradation. Even in the best space-based instruments, small amounts of material out gas, and after several weeks UV performance can be noticeably lower. One solution that seems to work, at least for the *Hubble Space Telescope* WFPC2 CCDs, is to warm the dewar to near 20°C for about 6 hours approximately every month.

Flat fields, as we have discussed before, are very important to have in one's calibration toolkit. Once in orbit, either as a satellite or a spacecraft destined for another world, the CCDs aboard generally have little ability to obtain flat field calibration images. High S/N flats made prior to launch in the laboratory are often the best available. These usually provide overall correction to 5% or a bit better, but small effects, such as illumination or instrument changes, limit the accuracy of the correction. Sometimes, the space-based CCD has significant changes, and large corrections are needed or new flats have to be generated in some manner.

The original WFPC camera aboard *Hubble* could obtain on-orbit flats through observation of the bright earth (Holtzman, 1990; Faber & Westphal, 1991). These were not elegant flats, having streaks and nonuniformities, but were all that was available. WFPC2 used Loral CCDs, which have an increased stability over the original TI CCDs, allowing preflight laboratory flats to work very well, even after the reduction in operating temperature. Numerous other small effects, such as color dependence, radiation damage, hot pixels, CCD illumination, and optical distortions, seen in the on-orbit WFPC2 flats are discussed in detail in Holtzman et al. (1995b). The effects of flat fielding, CTE, and the other issues discussed above on the photometric performance of the *Hubble* WFPC2 are described in Holtzman et al. (1995a), Whitmore & Heyer (1998), and Faber & Westphal (1991).

The *Galileo* spacecraft, currently providing impressive imagery of the planet Jupiter and its satellites, was one of the first public CCD cameras to be launched into space. Its CCD camera is described in detail in Belton et al. (1992) and can be used as an example of the details of space-based observations, their calibrations, properties, and difficulties. CCD and instrument stability and processes for their calibration after launch are major effects to consider as well as proper treatment of the photometric calibration images in lieu of the much reduced PSF.

The solid-state imager (SSI) aboard *Galileo* consisted of a single 800 × 800 TI CCD with a read noise of 40 electrons, gains of 38 to 380 electrons per DN, and a pixel size of 15 microns yielding 2.1 arcsec per pixel. The SSI, like the WFPC2, developed a CTE problem after about 8–12 months in space. Detailed study of SSI images taken during periods of cruise science (Howell & Merline, 1991) revealed that the CTE problem resulted in a readout tail containing 400 electrons, independent of the brightness of an imaged star or its location within the CCD. The cause was attributed to a trap, not in the active CCD array, but in the output register. Radiation damage (see next section) was the most likely cause. Owing to the constant number of trapped electrons, photometric correction was possible to a high degree of accuracy.

Point sources imaged in space are free from the blurring effects of the Earth's atmosphere and have a very small PSF compared with those commonly obtained with ground-based telescopes. A theoretical diffraction-limited image formed through a circular open aperture will have a FWHM (of the Airy disk) in radians of

$$FWHM = \frac{1.03\lambda}{D},$$

where λ is the wavelength of observation and D is the diameter of the aperture (Born & Wolf, 1959). Note that if we were to use the radius of the first Airy disk dark ring as our definition of image size, we would have the traditional formula

$$r = \frac{1.22\lambda}{D}.$$

Figure 7.2 shows theoretical Airy disk PSFs expected to be imaged by the SSI at three representative wavelengths and five different possible slight defocus values.

The FWHM of the SSI images (being obtained without any atmospheric or other seeing effects) were predicted to be about 0.55 arcsec at 4,000 Å and 1.2 arcsec at 9,000 Å. These PSF sizes correspond to 0.25 and 0.6 pixels respectively, making the SSI images severely undersampled ($r \sim 0.2$). This level of undersampling makes it impossible to directly determine the true FWHM or profile shape of a PSF. Using multiple images with slight offsets, images containing multiple stars with different pixel grid placements, and model CCD images, one can re-construct the true PSF imaged by an undersampled space-based CCD camera. In the SSI case, the PSF was found to be slightly larger than predicted and attributed to a slight camera focus problem.

As we have seen, undersampled images will lead to astrometric and photometric error, as the lack of a well-sampled PSF makes it hard to determine the true image center or the actual flux contained in the image. For the SSI, astrometric error amounted to about 0.8 arcsec even for bright stars, or about half a pixel. Observations of bright guide stars are a common occurrence for spacecraft and are used for navigation and course correction. Large astrometric uncertainties are hazardous and can lead to spacecraft orbital trajectories with inaccurate pointings, having the potential of producing spacecraft course corrections that could cause it to miss a target or, even worse, come too close. In the *Galileo* case, it was determined that a large number of guide star images were needed and careful analysis of these could be used to determine the path and navigation of the spacecraft within acceptable limits.

Photometrically, the nature of the undersampling manifests itself in two ways. First is the way in which one extracts the data and how a flux

Fig. 7.2. Modeled Airy disk patterns imaged by the *Galileo* SSI. The top panel shows the calculated PSFs as would be seen under very well-sampled conditions while the bottom panel shows the same PSFs as they would appear when imaged by the SSI. The severe pixelization of the PSFs is apparent. The rows are for 7,000, 5,500, and 4,000 Å (top to bottom) and the five columns are (left to right) defocus values for the SSI camera in mm. From Howell & Merline (1991).

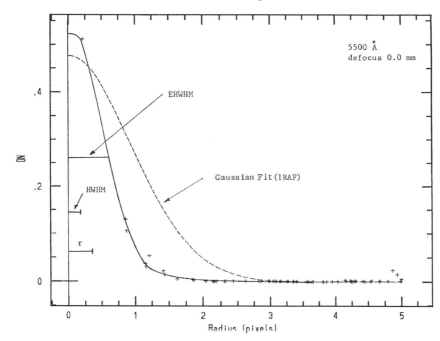

Fig. 7.3. Radial profile plot of a bright star imaged by the *Galileo* SSI. The plus signs are the actual CCD DN values (for $G = 380$ e$^-$/DN) and HWHM and r correspond to predicted values for an Airy disk imaged at 5,500 Å. Note that an approximation as a Gaussian profile is a poor representation of the actual PSF but the determined EHWHM for a single measurement is not far off. From Howell & Merline (1991).

value is assigned to it; second is the effect of digitization noise, which is large for the SSI. Figure 7.3 illustrates the first of these issues by presenting SSI data for a bright star. Because of the nature of the PSFs imaged with the SSI, one pixel (plus sign at $r = 0.25$) contains much more flux than any of the remaining ones. A standard Gaussian fit to these data (in this case made by *IRAF*) is seen to provide a complete misrepresentation of the image profile. Imagine the photometric error one would introduce by assumption of this type of profile and use of its shape as an indication of the total counts observed for this star. The effective FWHM (EFWHM) is defined as the apparent PSF width as determinable from a single undersampled image of a star. We see here that the EHWHM is 0.7 pixels, compared with the expected value (at 5,500 Å) of 0.55. The digitization effect present at the highest SSI gain setting leads to an uncertainty of ±379 electrons per DN. The above effects combined lead to an overall relative photometric uncertainty of 5–10% and an absolute spectrophotometric uncertainty of 10–30% for SSI data. These are higher than the 2–5% uncertainties quoted for the WFPC2 camera and are directly in proportion to the greater undersampling and higher CCD gain values used in the *Galileo* SSI.

Further readings concerning the special conditions and circumstances of CCDs when used for space-based observations can be found with a quick search of the websites of the *Hubble Space Telescope* and other satellite and spacecraft observatories. Access to numerous internal technical, engineering, and calibration reports is given as well as literature articles containing applications of the findings to astrophysical objects.

7.2 Radiation Damage in CCDs

With the launch of the *Galileo* spacecraft and the *Hubble Space Telescope*, astronomical imagery with CCDs from outer space began. With this exciting new windows on the universe came many unexpected effects in the performance and output noise levels of the CCDs involved. The study of radiation damage in CCDs had occurred in a number of military projects, but the low light and low noise levels needed for astronomy required new laboratory work and the development of techniques to deal with or avoid radiation effects altogether.

The hostile conditions expected in outer space were not the only radiation source to be concerned about for CCDs. Satellites in low Earth orbit, such as the *Hubble Space Telescope*, pass through the South Atlantic Anomaly (SAA) periodically, receiving healthy doses of high energy protons. *Galileo's* radioisotope thermal electric generator (RTG) provides power for the spacecraft as well as a neutron dose that bathes the onboard CCD imager. These inherent radiation environments, along with the general space background of cosmic rays and high energy particles from such events as solar flares or planetary magnetic fields, cause both temporary and permanent damage to a CCD in addition to long-term degradation.

Ironically, as CCDs became better astronomical devices in terms of their low read noise and dark currents, they also became much more susceptible to damage by high energy radiation. The SAA, for example, provides about 2,000 protons per square centimeter per second with energy of 50–100 MeV, for each passage. *Galileo's* RTG was expected to produce 10^{10} neutrons per square centimeter at the location of the CCD over the expected six year mission lifetime. Passage through Jupiter's radiation belts near the moon Io was predicted to provide a 2,500 rad dose of radiation to the CCD with each orbit. These levels of radiation do indeed cause damage to the CCD involved and methods of monitoring the changes that occur with time and the development of new manufacturing techniques aimed at radiation hardness were needed (McGrath, 1981).

The two major areas of concern in radiation damage to CCDs are (1.) high energy photon interactions, which result in fast electrons, which in turn cause simple, localized damage defects and the generation of numerous electron–hole pairs, and (2.) nuclear reactions caused by uncharged neutrons or high energy protons, which cause large area defects and are more likely to lead to partial or complete failure of a device (Janesick, Elliott, & Pool, 1988). The first of these radiation induced concerns is called an ionization effect and involves gamma rays or charged particles. The second, involving massive particles, is termed a bulk effect or displacement damage owing to its ability to displace silicon atoms from their lattice positions within the CCD.

Displacement damage can involve single silicon atoms or bulk damage involving clusters of atoms, all removed from their original lattice locations within the CCD. The vacancies remaining in the lattice structure create trapping locations, which in turn cause degraded or no CTE performance for one or more pixels in the array. As the result of lattice stresses, the trap locations become populated by one or more of the doping elements such as phosphorus. The presence of a phosphorus atom within the silicon lattice modifies the band gap energies locally and is thought to be the cause of the observed reduced CTE effects (Srour, Hartmann, & Kitazaki, 1986).

Repair of single lattice defects has been accomplished by heating the CCD (to room temperature or higher), while bulk defects are essentially impossible to repair. It has been noticed, however, that at low temperatures ($< -100°$C), the trapped charge accumulated at the defect location remains trapped and has little effect on the overall CTE. This temperature dependence has been shown to be proportional to $\exp(-E_T/kT)$, where E_T is the activation energy of the lattice traps (Janesick, Elliott, & Pool, 1988). Thus, one way to avoid lattice defects is to operate the CCD at temperatures as low as possible. Interestingly, high temperature operation ($>30°$C) allows trapped charge to be released very quickly, eliminating a deferred charge tail and providing good CTE. These various techniques involving temperature manipulation of a CCD system are hard to employ in space. The temporal behavior of the CCD involved can be unpredictable and may be different for each CCD, even those of the same type.

Ionization effects, caused by gamma rays or charged particles, cause a charge buildup in the CCD gate structures and can produce increases in the CCD dark current. Whereas a 150 keV electron is needed to cause an actual silicon atom displacement, only a few eV of energy deposited in the gate insulator is enough to change the potential and cause charge trapping. Even an intense UV flood (Schaeffer et al., 1990; Ditsler, 1990) with 2,500 Å photons can cause dark current increases or even render the CCD inoperable. The charge buildup causes new states to exist within the band gap of the silicon leading to easier generation of thermal electrons and thereby an increased dark current. The affected CCD pixels, that is, those which have been damaged by the ionizing radiation, will show increased dark current, while their neighbors will not. Histograms of the amount of dark current produced as a function of pixel signal level often show "spikes" of dark current at specific signal levels. This is taken to indicate a sort of quantized structure in the amount of damage that occurs per radiation site (Janesick, Elliott, & Pool, 1988).

Methods of protecting CCDs from radiation effects are varied. The type and amount of radiation expected and the scientific goals of the imager must be carefully weighed to produce the final compromise. For example, the CCD flown on the *Galileo* mission was initially tested for its ability to withstand gamma radiation similar to that expected in the Jovian magnetic fields, a test which it passed well. It was probably by shear luck that a test or two was also performed to understand its performance when exposed to neutrons. Increased dark current was prevalent and, without time and money to redesign the imager, operation at a temperature of $-130°$C was implemented compared with the original proposed value of $-40°$C. In the case of the *Hubble Space Telescope* CCDs, the increased dark current was not a large factor because of their initially colder operating temperature, but long-term degrading CTE and QE effects were noticed (Holtzman, 1990) and attributed to in-orbit radiation damage. A detailed report of the detectors in the *Hubble* telescope is contained in Brown (1993) and considerations on improving the ability of CCDs to counteract the effects of radiation are discussed in IEEE Nuclear Science Symposium (1988) and Bely, Burrows, & Illingworth (1989).

7.3 CCDs in the UV and EUV (300–3,000 Å) Spectral Range

Progress in the use of CCDs for UV and EUV observations has occurred on two main fronts. Coatings applied to the CCDs to down-convert high energy photons to visible light photons is one method. The other method involves new manufacturing techniques that allow short wavelength photons to penetrate the silicon even given the very small absorption depths at these wavelengths.

Coating the CCD with a UV phosphor has been discussed previously in this book (Chapter 2) with regard to enhancement of the near-UV wavelength range. These same coatings can often increase the QE of a CCD to usable levels, down to wavelengths as short as 500 Å. Lumogen, a UV phosphor that absorbs photons of $\lambda < 4,200$ Å and reemits their energy near 5,200 Å, is a popular choice. This inexpensive material is thermally deposited onto the CCD surface while under vacuum into a layer about 6,000 Å thick. Use of Lumogen as a coating delivers a QE from 500–4,000 Å of around 15% (yielding about 3% with a solar blind filter in place) and actually increases the intrinsic QE from 5,000–8,000 Å, as it acts as an antireflection coating. Phosphor coatings can be deposited onto either front- or back-side illuminated CCDs (McLean, 1997c). Other coatings such as coronene and Metachrome II are also used for UV and EUV enhancement (Trauger, 1990; Geary et al., 1990; Schempp, 1990).

Modern manufacturing processes have again come to the rescue through the development of techniques that allow the CCD itself to have more sensitivity to short wavelength photons. We have mentioned before that the gate structures of a CCD can absorb short wavelength photons before they enter the actual pixel itself, thus reducing or eliminating the collection probability at these wavelengths. Solutions to this problem consisted of using a back-side illuminated, thinned device or making CCDs with transparent gates. Both of these techniques can be further improved and employed for detection of UV and shorter wavelength photons.

During the thinning process, the back side of a CCD forms an oxide layer, which can include a surface layer of incomplete bonds that are positively charged. With the very short absorption depths for UV and shorter wavelength photons, any photoelectrons produced within the silicon are more likely to be attracted to the positive bonds where recombination occurs, and not to the pixel potential well for collection, storage, and output. Various techniques have been developed to reduce the number of positive incomplete bond sites that exist in a thinned CCD (Janesick et al., 1985; Janesick et al., 1989; Baily et al., 1990). Additionally, methods that allow removal of the oxide layer produced in the thinning process have been developed and consist of precision etching of the oxide layer under controlled conditions (Bonanna, 1995).

Recent detailed work, aimed at understanding the characteristics and performance of CCDs in the wavelength range of 300–4,000 Å, has been performed. The laboratory setup used and various types of short wavelength enhanced CCDs produced and tested are described in Bonanna (1995), which concludes that QE values of 10–60% can be achieved over the wavelength range of 300–2,500 Å. These same CCDs also have up to 80% QE at 6,000 Å. Phosphor coatings and manufacturing improvements are about equal in their ability to enhance UV and EUV performance; however, both types of improvement appear to show a decrease in their QE with time, once the CCD is cooled and put under vacuum. Contamination by outgassing within the vacuum and subsequent freezing of the contaminants onto the CCD surface are thought to be the most likely cause of the reduced QE.

To overcome the low QE of CCDs at short wavelengths, even after the above enhancements have been performed, some high-energy applications employ standard unmodified CCDs as detectors, preceded in the optical path by a device such as a microchannel plate (MCP). MCPs can produce up to about 500,000 electrons per incident high energy photon, and this electron cloud strikes a phosphor coated photocathode producing visible light photons that are collected and imaged by the CCD (Eccles, Sim, & Tritton, 1983). MCPs operate at high voltages (a few keV) and are inherently solar blind as they require high energy photons for activation. Final QE values of up to 20% are possible with a well-constructed device. This increased QE is the largest advantage of instruments that use intensified CCDs, while poor spatial resolution, phosphor decay effects, and smaller dynamic range (compared with a normal CCD) are the major disadvantages (McLean, 1997c; Longair, 1997).

7.4 CCDs in the X-Ray (<500 Å) Spectral Range

Figure 7.1 provided hints that CCDs may also be useful detectors for the X-ray region of the spectrum, as the absorption depth within silicon rises shortward of about 1,000 Å. Figure 7.4 shows us a similar result, only this time we express it as the quantum efficiency of the CCD as a function of photon energy or wavelength. We note that within the X-ray region, back-side thinned CCDs are extremely efficient detectors, approaching a quantum efficiency of 100% at times. Currently planned X-ray telescopes such as *XMM* and *AXAF* are using CCDs as their detectors (Longair, 1997).

X-ray detection by CCDs works in a slightly different manner than detection of optical photons. An incident optical photon creates a photoelectron within the silicon lattice, which moves from the valance to

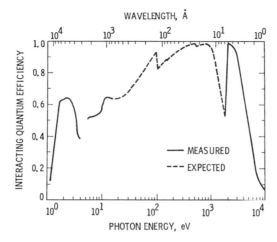

Fig. 7.4. Quantum efficiency for a typical thinned, back-side illuminated CCD from the X-ray to the optical spectral regions. From Janesick et al. (1988).

the conduction band and is then held there (in a pixel) by an applied potential. The absorption of an X-ray photon by silicon ejects a free photoelectron of energy $E - b$, where $E = h\nu$ and b is the binding energy of the electron to the silicon atom, typically 1,780 eV. As this free electron moves through the silicon lattice, it produces a trail of electron–hole (e–h) pairs, each requiring an average of 3.65 eV of energy to be produced. If all the energy of the free electron went into e–h pair production, the energy of the incident X-ray could be precisely knowable simply by counting the ADUs produced within the CCD pixels. However, a small undetermined amount of the free electron's energy goes into various phononic states of the silicon lattice, thereby causing some uncertainty in the value of the incident photon's energy. The level of this uncertainty, the "Fano" factor,[†] is so small that to obtain Fano-noise limited CCD performance, the CCD read noise must be less than about 2 electrons (Janesick et al., 1988).

There are two very interesting characteristics of using CCDs for X-ray detection. One, just alluded to above, is that X-ray imaging with a CCD provides not only an image but also the ability to produce a spectrum from the image. Each incident X-ray photon collected produces a measureable number of e–h pairs, thus yielding a method by which one

---

[†] The term Fano factor is due to U. Fano who, in 1947, formulated a description of the uncertainty in the energy of ion pairs produced in a gas by ionizing radiation.

can backtrack and obtain the incident photon energy (Longair, 1997). X-ray detection can also be used as an excellent diagnostic tool of the CTE of a CCD. Using, for example, a [55]Fe source (which produces X-ray photons due to the 5.9 keV Kα Fe line), each absorbed X-ray photon will deposit 1,620 electrons into each pixel. Acquisition and examination of a [55]Fe CCD image can help discover problems with the device (Janesick et al., 1988; Janesick et al., 1987b). CCD pixel defects or problems with the associated electronics manifest themselves as pixels in the array with less than 1,620 collected electrons, smeared charge trails showing poor CTE, and other visible diagnostic events.

The use of CCDs as detectors for X-ray imaging and spectroscopy is still in an early phase. This chapter has provided a very brief introduction to the use of CCDs for detection and study of photons with energies higher then those in the optical wavelength region. The literature is just beginning to grow with articles related to high energy uses of CCDs in astronomical observations. For example, the laboratory work involved in the realization of the CCD detectors for the *AXAF* and *XMM* satellite missions has involved the development of new technologies and new analysis procedures. The interested reader is referred to the recent works by McLean, Longair, and others listed in Appendix A.

A

CCD Reading List

This appendix provides a reading list covering the aspects of CCD devel opment, research, and astronomical usage. There are so many articles, books, and journal papers covering the innumerable aspects of information on CCDs that the material presented in a book this size or any size can only cover a small fraction of the details of such work. Even the list presented here does not cover all apsects of interest concerning the use of CCDs in astronomy, but it does provide a very good starting point. The growth of information on CCDs has risen sharply over the past ten years and will, no doubt, continue to do so. Thus, the student of CCD science must constantly try to keep up with the latest developments both in astronomy and within the field of opto-electronics, both areas where progesss is being made.

This appendix lists the reading material by broad catagory, but keep in mind that many works contain information on specific charateristics of certain types of CCD as well as information as to their use and problems. Scientific journal articles are not listed here since they are referenced within the papers presented, are provided throughout the text, and are easily searched for using the journal indexing systems provided in printed and web based forms.

Numerous nontechnical articles related to CCDs have appeared in such magazines as *Astronomy*, *Sky & Telescope*, *Scientific American*, and *CCD Astronomy*.[†] The reader is cautioned that in some of these descriptions of CCDs, the details are not always completely correct nor have the methodologies always been rigorously tested. However, some of the "popular" articles contain a wealth of information that would be

[†] *CCD Astronomy* is no longer published but has been incorporated into *Sky & Telescope*.

unlikely to appear elsewhere. We refer the reader to the websites of these magazines for a list of relevant articles.

Every major observatory is a storehouse of knowledge on CCDs. They have websites filled with useful information, large numbers of engineering and technical reports covering aspects of instrument design and construction, manuals for the observatory instruments available, and often newsletters that contain very worthwhile information on their particular CCDs. Many laboratories working on CCDs also have websites that are extremely valuable (see Appendix B).

A.1 General CCD References

Listed below are some general references containing details on CCDs. They are listed by year of publication, and the end of the list contains yearly or periodic references.

Eccles, M. J., Sim, M. E., & Tritton, K. P., 1983, *Low Light Level Detectors in Astronomy*, Cambridge University Press.

Borucki, W. & Young, A., eds., 1984, *Proceedings of the Workshop on Improvements to Photometry*, NASA Conference Publication 2350.

Dereniak, E. & Crowe, D., 1984, *Optical Radiation Detectors*, Wiley.

Hearnshaw, J. B. & Cottrell, P. L., eds., 1986, *Instrumentation and Research Programmes for Small Telescopes*, D. Reidel.

Walker, G., 1987, *Astronomical Observations*, Cambridge University Press.

Borucki, W., ed., 1988, *Second Workshop on Improvements to Photometry*, NASA Conference Publication 10015.

Robinson, L. B., ed., 1988, *Instrumentation for Ground-Based Optical Astronomy*, Springer-Verlag.

Jacoby, G. H., ed., 1990, *CCDs in Astronomy*, ASP Conference Series Vol. 8.

Philip, A. G. D., Hayes, D., & Adelman, A., eds., 1990, *CCDs in Astronomy II: New Methods and Applications of CCD Technology*, L. Davis Press.

Howell, S. B., ed., 1992, *Astronomical CCD Observing and Reduction Techniques*, ASP Conference Series Vol. 23.

Butler, C. J. & Elliot, I., eds., 1993, *Stellar Photometry – Current*

Techniques and Future Developments, IAU Colloquium 136, Cambridge University Press.

Kilkeny, D., Lastovica, E., & Menzies, J. W., eds., 1993, *Precision Photometry*, South African Astronomical Observatory Publication.

Sze, S. M., 1994, *Semiconductor Sensors*, Wiley.

Philip, A. G. D., Janes, K., & Upgren, A., eds., 1995, *New Developments in Array Technology and Applications*, IAU Symposium No. 167, Kluwer.

Reike, G., 1996, *The Detection of Light from Ultraviolet to the Submillimeter*, Cambridge University Press.

McLean, I. S., 1997, *Electronic Imaging in Astronomy*, Wiley.

Longair, M., 1997, *High Energy Astrophysics*, Vols. 1 & 2, Cambridge University Press.

Holst, G. C., 1998, *CCD Arrays, Cameras, and Displays*, SPIE Press.

Beletic, J. W. & Amico, P., eds., 1998, *Optical Detectors for Astronomy*, Kluwer.

Martinez, P. & Klotz, A., 1998, *A Practical Guide to CCD Astronomy*, Cambridge University Press.

Modern Technology and Its Influence on Astronomy, Chapters 20 and 21, J. V. Wall & A. Boksenberg, eds., Cambridge University Press.

Photonics Handbook. Yearly editions available from Laurin Publishing Company Inc., Pittsfield, MA.

Optical Engineering. The primary journal of the International Society for Optical Engineers, has numerous special editions devoted to CCDs.

Proceedings of SPIE. The International Society for Optical Engineering. Numerous volumes over the past twenty years or so. Of special interest are the volumes on "Instrumentation in Astronomy," and "Advanced Technology Optical Telescopes."

B
CCD Manufacturers and CCD Website Information

Table B.1 lists present day CCD manufacturers and this appendix lists internet sites of interest to CCD users. In addition to the companies that make CCDs, a number of other internet locations provide detailed information of interest. These are laboratories that study, characterize, and modify CCDs for astronomical use, manufacturers of CCDs and complete systems, and software houses that specialize in CCD codes. The list below is not exhaustive but does provide a good jumping off point. A search of the internet using "CCD" will provide at least an afternoon of reading material and numerous other sites to visit.

Solid State Devices Group at the University of Leicester
    http://www.star.le.ac.uk/solidstate/ssdg.html
University of Arizona, Steward Observatory, CCD Lab
    http://sauron.as.arizona.edu/ccdlab/
CCD Section of the Electronic Universe Web Project
    http://zebu.uoregon.edu/ccd.html
San Diego State University CCD Lab
    http://mintaka.sdsu.edu/ccdlab/LabMain.html
University of Hawaii CCD Lab
    http://ccd.ifa.hawaii.edu/
MIT CCD X-Ray Group
    http://space.mit.edu/ACIS/Welcome.html
European Southern Observatory
    http://www.hq.eso.org/odteam/generalccd/generalccd.html
National Optical Astronomy Observatories (KPNO & CTIO)
    http://www.noao.edu/

Hubble Space Telescope
  http://www.stsci.edu/
Royal Observatory
  http://www.roe.ac.uk/
South African Astronomical Observatory
  http://www.saao.ac.za/
Anglo-Australian Observatory CCD Imaging manual
  http://www.aao.gov.au/local/www/cgt/ccdimguide/manula.html
Joint Astronomy Center in Hawaii
  http://www.jach.hawaii.edu
UCO/Lick Detector lab
  http://gardiner.ucolick.org/ccdev/ccddist.html
CFHT CCD Selection
  http://www.cfht.hawaii.edu/manuals/detectors/CFHT_CCD.html
In Situ Tests for CCDs at the Telescope
  http://www.cfht.hawaii.edu/tmca/top/top.html
Photometrics Corp.
  http://www.photomet.com/
Scientific Imaging Technologies (SITe). CCD Tutorial
  http://www.site-inc.com/tutorial.htm
Royal Greenwich Observatory CCD Page
  http://www.ast.cam.ac.uk/apo/docs/ccds.html
CCDs at the La Palma Observatories
  http://www.ing.iac.es (see information for users)
Apogee Instruments Inc.
  http://www.apogee-ccd.com/
PixelVision Inc.
  http://www.pv-inc.com/
Princton Instruments
  http://www.prinst.com/
SpectraSource Instruments
  http://www.optics.org/spectrasource/
Observatory Techniques Magazine (amateur)
  http://home.midco.net/otm/OTM.HTML
Osservatorio Astrofisico di Catania, Istituto di Astronomia
  http://www.ct.astro.it/
Osservatorio Astrofisico di Catania, Istituto di Astronomia: CCD Lab
  http://www.ct.astro.it/ccdlab/index.html

Table B.1. *CCD detector manufacturers*

Company	Company Location	CCD Name	CCD Types
Dalsa Inc.	Ontario, Canada	Various	Large formats; small pixels
Eastman Kodak Co.	Rochester, NY, USA	Kodak	Large formats; small pixels
EG&G Reticon	Sunnyvale, CA, USA	EG&G	Various
English Electric Valve	Essex, England	EEV	Scientific; X-ray
Loral/Fairchild Imaging Sensors	Newport Beach, CA, USA	Fairchild/Ford	Large formats
MIT/Lincoln Labs	Lexington, MA, USA	Lincoln Labs	High speed read out; low noise
Orbit Semiconductor	Sunnyvale, CA, USA	Orbit	Very large formats
Philips	Eindhoven, Netherlands	Philips	Unique CCDs, new products
David Sarnoff Research Center	Princton, NJ, USA	Various	Various
Scientific Imaging Technologies	Beaverton, OR, USA	SITe	Various
Texas Instruments	Dallas, TX, USA	TI	Various
Thomson	Cedex, France	Thomson	Various; large formats

Table B.1. *(Continued)*

Internet Web Address	Comments
http://www.dalsa.com/	Designs CCD systems; fabricates CCDs via foundaries
http://www.kodak.com/	Wide range of CCD types; interline and frame transfer devices
—	Developed large format arrays with Lick Observatory
http://cco-web.com/register/00281.html	High performance scientific devices; close ties to astronomy; foundry runs
—	Close ties with astronomy; foundry runs; Fairchild + Ford Aeronutronics
http://www.ll.mit.edu/	Federally funded R&D facility, not publically available
http://www.eedesign.com/VendorGuide/Orbit.html	CMOS and CCD foundaries; produce CCD foundry runs
http://207.87.19.21/	CCDs produced for internal use
http://www.sarnoff.com/	Former RCA Labs; contract work
http://www.site inc.com	Former Tektronix imaging group; foundry service
http://www.ti.com/sc/docs/msp/ccd.htm	No longer produces scientific grade CCDs, industrial and consumer varieties
http://www.esi.es/About esi/Members/thomson.html	Scientific & commercial CCDs; close ties with ESO

C

Some Basics of Image Displays

Most computer screens and image displays in use are 8-bit devices. This means that the displays can represent data projected on them with $2^8 = 256$ different greyscale levels or data values of resolution. These greyscale levels can represent numeric values from 0 to 255 and it is common to only have about 200 levels actually available to the image display for representing data values with the remaining 50 or so values reserved for graphical overlays, annotation, etc. If displaying in color (actually pseudo-color), then one also has available about 200 separate colors, each with a possible grey value of 0–255, or the famous "16 million possible colors" listed in many computer ads.

On the display, the color black is represented by a value of zero (or in color by a value of zero for each of the three color guns, red (R), green (G), and blue (B)). White has R = G = B = 255, and various grey levels are produced by a combination of R = G = B = N, where N is a value from 0 to 255. Colors are made by having R $\neq$ G $\neq$ B or any combination thereof in which all three color guns are not operated at the same intensity. A pure color, say blue, is made with R = G = 0 and B = 255 and so on. You may have noticed that color printers have three (or four) colors of ink in them. They contain cyan, blue, and magenta (and black) inks, which are used in combination to form all the output colors. This difference (cyan etc. vs. RGB) in the choice of colors is simply because display screens mix light whereas printers mix ink to form specific colors.

Terms one hears but rarely uses in astronomy are hue, saturation, and brightness. Hue means the color of the image, saturation is the relative strength of a certain color (fully saturated = 1), and brightness is the total intensity of a color where black = 0. When you change colors (RGB) you are really changing the hue, saturation, and brightness of the image display. These three terms are fully explored in Gonzalez & Woods (1993) as well as in almost any text introducing image processing techniques.

Almost all CCD data obtained today have a dynamic range of much greater then 8 bits. Thus, in order to display the CCD image, some form of scaling must be performed to allow the image to be shown on a display with only 8 bits. A common technique (often performed by the software without user intervention) is called linear scaling. This type of scaling divides the entire true data range into say 200 equal bins, where each bin of data is represented by 1 of the 0–200 available greyscale levels. In a pseudo-color representation, the 200 levels are further divided into three bins, one for each color. For example, if an image has real data values in the range from 0 to 100,000 ADUs, linear scaling will place the real data values between 0 and 500 ADU into the first scaled bin and will display them as a 0 on the screen. If your image is such that all the interesting astronomical information have real values of 0 to 2000, this linear scaling scheme will represent all the real image information for the values of 0 to 2000 within only 4 display bins, those having values of 0–1.

To avoid such poor scaling and loss of visual information, two alternatives generally exist: Either one uses a linear scaling but within a specific data window or one uses a different type of scaling altogether. The first option is accomplished by having the software again perform a linear scaling but this time using its 200 output display levels to scale image data values only within the data window of say 0 to 2000. Different scaling options allow for nonlinear modes such as log scaling, exponential scaling, histogram equalization, and many others. These are easily explored in any of the numerous image processing software packages used today

A method commonly used to aid the eye when viewing a displayed image is that of interactive greyscale manipulation. You probably know this as changing the image stretch or contrast and perform it via movement of a mouse or trackball while displaying an image. The actual change that is occurring is a modification of the relation between the input data values to those output to the display screen. The software mechanism that controls this is called a look-up table or LUT. Some sample LUTs are shown in Figure C.1, where we see the relation of input to output data value. For greyscale images, all three color LUTs are moved in parallel, while in pseudo-color mode, each color LUT can be individually controlled. Changes in the slope and intercept of the LUT

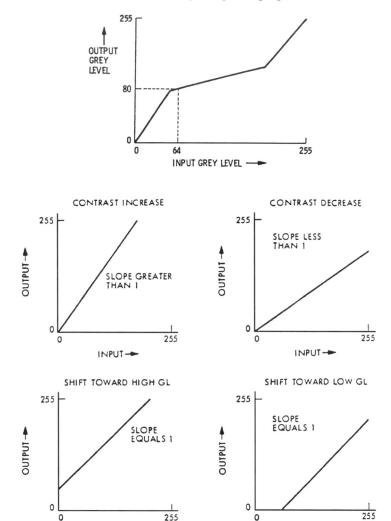

Fig. C.1. The top panel shows an example LUT, which converts input pixel values or grey levels to output grey levels on the video display. In this particular case, an input pixel value of 64 will be displayed with a grey level value of 80. Note that the relationship between input and output is not a single-valued linear transformation. The bottom panel shows various linear stretch operations. If the mapping passes through the origin and has a slope >1, the effect will be to increase the output contrast. A slope of <1 decreases the contrast. If the linear mapping intersects the y axis (i.e., the output value axis) the input values will be systematically shifted to higher output grey levels. Intersecting the horizontal axis causes a net shift to lower output grey levels. From Gonzalez & Woods (1993).

transfer function control changes in the image brightness, contrast, and color. Terms such as "linear stretch" simply refer to a LUT using a linear transformation function.

In the above discussion, we have discussed pseudo-color, but what about true color? True color attempts to display imagery in a manner that is similar to how our eye might perceive a scene. In the visible bandpass, this amounts to trying to match the RGB response of our eye by displaying three separately scaled 8-bit images. Thus, true color displays must have a way to display three separate 8-bit images at once. This is done using a 24-bit display consisting of three individual 8-bit image planes. For true color representation of data from outside the range of the human eye, the user must approximate what the scene might look like in each of the three "colors." The user or display software creates some method of scaling the nonvisible data to the RGB values within the display device itself.

Further information on general image processing techniques can be found in Gonzalez & Woods (1993), while additional details on image displays can be found in Hanisch (1992). User manuals for astronomical image processing software packages such as IRAF and MIDAS often have numerous examples and routines useful for dealing with CCD imagery.

References

Adams, M., Christian, C., Mould, J., Stryker, L., & Tody, D., 1980, *Stellar Magnitudes from Digital Pictures*, Kitt Peak National Observatory publication.

Amelio, G., Tompsett, M., & Smith, G., 1970, *Bell Systems Technical Journal*, **49**, 593.

Baily, P., et al., 1990, *Proc. SPIE*, **1344**, 356.

Baum, W. A., Thomsen, B., & Kreidl, T. J., 1981, in *Solid State Imagers for Astronomy*, eds. J. C. Geary & D. W. Latham, *Proc. SPIE*, **290**, 24.

Belton, M., et al., 1992, *Space Sci. Rev.*, **60**, 413.

Bely, P.-Y., Burrows, C., & Illingworth, G. (eds.), 1989, *The Next Generation Space Telescope*, Space Telescope Science Institute publication.

Bland, J. & Tully, R., 1989, *Astron. J.*, **98**, 723.

Blouke, M., Yang, F., Heidtmann, D., & Janesick, J., 1988 in *Instrumentation for Ground-Based Optical Astronomy*, ed. L. B. Robinson, Springer-Verlag, p. 462.

Bonanna, G., 1995, in *New Developments in Array Technology and Applications*, eds. A. G. D. Philip, K. A. Janes, & A. R. Upgren, Kluwer, p. 39.

Born, M. & Wolf, E., 1959, *Principles of Optics*, MacMillan, Chap. VIII.

Boulade, O., Vigroux, L., Charlot, X., Borgeaud, P., Carton, P., de Kat, J., Rousse, J., & Mellier, Y., 1998, in *Optical Astronomical Instrumentation*, ed. S. D'Odorico, *Proc. SPIE*, **3355**, 614.

Bowen, I. S., 1960a, in *Astronomical Techniques*, ed. W. A. Hiltner, University of Chicago Press, Chap. 2.

Bowen, I. S., 1960b, in *Telescopes*, eds. G. Kuiper & B. Middlehurst, University of Chicago Press, p. 1.

Boyle, W. & Smith, G., 1970, *Bell Systems Technical Journal*, **49**, 587.

Brar, A., 1984, *AURA Engineering Technical Report No. 76*.

Broadfoot, A. L. & Kendall, K., 1968, *Journal of Geophys Res. Space Physics*, **73**, 426.

Brown, R. (ed.), 1993, *The Future of Space Imaging*, Space Telescope Sciene Institute publication, Chap 8.

Buonanno, R. & Iannicola, G., 1989, *Publ. Astron. Soc. Pac.*, **101**, 294.

Burke, B., Reich, R., Savoye, E., & Tonry, J., 1994, *IEEE Trans. Electron Devices*, **41**, 2482.

Chiu, L.-T., 1977, *Astron. J.*, **82**, 842.

Chromey, F. & Hasselbacher, D. A., 1996, *Publ. Astron. Soc. Pac.*, **108**, 944.

Cochran, A., 1995, in *New Developments in Array Technology and Applications*, eds. A. G. D. Philip, K. A. Janes, & A. R. Upgren, Kluwer, p. 251.

Corbally, C., 1995, in *New Developments in Array Technology and Applications*, eds. A. G. D. Philip, K. A. Janes, & A. R. Upgren, Kluwer, p. 241.

Dacosta, G., 1992, in *Astronomical CCD Observing and Reduction Techniques*, ASP Conference Series Vol. 23, ed. S. Howell, p. 90.

Davis, L. E., 1994, *A Reference Guide to the IRAF/DAOPHOT Package*, NOAO publication.

Delamere, A., Atkinson, M., Rice, J., Blouke, M., & Reed, R., 1990, in *CCDs in Astronomy*, ASP Conference Series Vol. 8, ed. G. H. Jacoby, p. 269.

DeVeny, J., 1990, *The Gold Camera Instrument Manual: A CCD Spectrograph for the 2.1 Meter Telescope*, NOAO Instrument Manual Publication.

Dhillon, V., Rutten, R., & Jorden, P. R., 1993, *Gemini*, Issue 41, p. 7.

Diego, F., 1985, *Publ. Astron. Soc. Pac.*, **97**, 1209.

Ditsler, W., 1990, in *CCDs in Astronomy*, ASP Conference Series Vol. 8, ed. G. H. Jacoby, p. 126.

Djorgovski, S., 1984, in *Proceedings of the Workshop on Improvements to Photometry*, eds. W. J. Borucki & A. Young, NASA Conf. Publ. 2350.

Eccles, M., Sim, E., & Tritton, K., 1983, *Low Light Level Detectors in Astronomy*, Cambridge University Press.

Faber, S. & Westphal, J. (eds.), 1991, *WFPC Science Verification Report*, Space Telescope Science Institute.

Fehrenbach, C., 1967, in IAU Symposium 30, eds. A. Batten & D. Heard, Academic Press, p. 65.

Florentin-Nielsen, R., Anderson, M., & Nielsen, S., 1995, in *New Developments in Array Technology and Applications*, eds. A. G. D. Philip, K. A. Janes, & A. R. Upgren, Kluwer, p. 207.

Furenlid, I. & Meylon, T., 1990, in *CCDs in Astronomy II: New Methods and Applications of CCD Technology*, eds. A. G. D. Philip, D. Hayes, & S. Adelman, L. Davis Press, p. 13.

Geary, J., Torres, G., Latham, D., & Wyatt, W., 1990, in *CCDs in Astronomy*, ASP Conference Series Vol. 8, ed. G. H. Jacoby, p. 40.

Gehrels, T., Marsden, B., McMillian, R., & Scotti, J., 1986, *Astron. J.*, **91**, 1242.

Gibson, B., 1991, *J. R. Astron. Soc. Canada*, **85**, 158.

Gibson, B. & Hickson, P., 1992, *Mon. Not. R. Astron. Soc.*, **258**, 543.

Gilliland, R., 1992, in *Astronomical CCD Observing and Reduction Techniques*, ASP Conference Series Vol. 23, ed. S. Howell, p. 68.

Gilliland, R. L., et al., 1993, *Astron. J.*, **106**, 2441.

Gonzalez, R. & Woods, R., 1993, *Digital Image Processing*, Addison-Wesley.

Gorjian, V., Wright, E., & Mclean, I., 1997, *Publ. Astron. Soc. Pac.*, **109**, 821.

Gudehus, D., 1990, in *CCDs in Astronomy*, ASP Conference Series Vol. 8, ed. G. H. Jacoby, p. 356.

Gullixson, C., 1992, in *Astronomical CCD Observing and Reduction Techniques*, ASP Conference Series Vol. 23, ed. S. Howell, p. 130.

Gunn, J. & Westphal, J., 1981, *Proc. SPIE*, **290**, 16.

Gunn, J., et al., 1998, *Astron. J.*, **116**, 3040.

Hanisch, R., 1992, in *Astronomical CCD Observing and Reduction Techniques*, ed. S. B. Howell, ASP Conf. Series Vol. 23, p. 285.

Hayes, D. S. & Latham, D. W., 1975, *Astrophys. J.*, **197**, 593.

Hendon, A. & Kaitchuck, R., 1982, *Astronomical Photometry*, Van Nostrand Reinhold.

Hickson, P. & Richardson, E. H., 1998, *Publ. Astron. Soc. Pac.*, **110**, 1081.

Hickson, P., Borra, E., Cabanac, R., Content, R., Gibson, B., & Walker, G., 1994, *Astrophys. J. Lett.*, **436**, L201.

Hiltner, W. A. (ed.), 1962, *Astronomical Techniques*, Vol. II of *Stars and Stellar Systems*, University of Chicago Press.

Holtzman, J. A., 1990, *Publ. Astron. Soc. Pac.*, **102**, 806.

Holtzman, J., et al., 1995a, *Publ. Astron. Soc. Pac.*, **107**, 1065.

Holtzman, J., et al., 1995b, *Publ. Astron. Soc. Pac.*, **107**, 156.

Honeycutt, R. K., 1992, *Publ. Astron. Soc. Pac.*, **104**, 435.

Horne, K., 1986, *Publ. Astron. Soc. Pac.*, **98**, 609.

Howell, S. B., 1989, *Publ. Astron. Soc. Pac.*, **101**, 616.

Howell, S. B., 1992, in *Astronomical CCD Observing and Reduction Techniques*, ASP Conference Series Vol. 23, ed. S. Howell, p. 105.

Howell, S. B., 1993, in *Stellar Photometry – Current Techniques and Future Developments*, eds. C. J. Butler & I. Elliott, IAU Colloquium 136, Cambridge University Press, p. 318.

Howell, S. D., Everett, M., Esquerdo, G., David, D., Weidenschilling, S., & Van Lew, T., 1999, in *Precision CCD Photometry*, eds. E. Craine & D. Crawford, ASP Conference Series, Vol. 189.

Howell, S. B., Everett, M., & Ousley, D., 1999, in *Third Workshop on Improvements to Photometry*, ed. W. Borucki, NASA Conference Publication, in press.

Howell, S. B. & Jacoby, G., 1986, *Publ. Astron. Soc. Pac.*, **98**, 802.

Howell, S. B., Koehn, B., Bowell, E. L. G., & Hoffman, M., 1996, *Astron. J.*, **112**, 1302.

Howell, S. B. & Merline, W., 1991, "Results from the November 1990 Galileo Satellite SSI Calibration," *Memorandum to Galileo SSI Team*, 8 July 1991 & Addenda A (19 Aug. 1991) and B (5 Oct. 1991), Galileo SSI Team memoranda, Galileo Satellite Project, NASA/JPL.

Howell, S. B., Mitchell, K. J., & Warnock, A, III, 1988, *Astron. J.*, **95**, 247.

Howell, S. B. & Usher, P. D., 1993, *Publ. Astron. Soc. Pac.*, **105**, 383.

IEEE Nuclear Science Symposium, 1988, "Radiation Damage in Scientific Charge-Coupled Devices," *IEEE Trans.*, 42.

Jacoby, G., Hunter, D., & Christian, C., 1984, *ApJ Suppl.*, **56**, 257.

Janesick, J. & Blouke, M., 1987, *Sky and Telescope Magazine*, September, **74**, 238.

Janesick, J. & Elliott, T., 1992, in *Astronomical CCD Observing and Reduction Techniques*, ASP Conference Series Vol. 23, ed. S. Howell, p. 1.

Janesick, J., Elliott, T., Bredthauer, R., Chandler, C., & Burke, B., 1988, *Optical and Optoelectronic Applied Science and Engineering Symposium; X-ray Instrumentation in Astronomy*, SPIE, Bellingham, WA.

Janesick, J., Elliot, T., Collins, M., Blouke, M., & Freeman, J., 1987a, *Opt. Eng.*, **26**, 692.

Janesick, J., Elliot, T., Collins, S., Daud, T., Campbell, D., & Garmire, G., 1987b, *Opt. Eng.*, **26**, No. 2.

Janesick, J., Elliot, T., Daud, T., McCarthy, J., & Blouke, M., 1985, in *Solid State Imaging Arrays*, eds. K. N. Prettyjohns & E. L. Dereniak, *Proc. SPIE*, **570**, 46.

Janesick, J., Elliot, T., Fraschetti, G., Collins, S., Blouke, M., & Corrie, B., 1989, *Proc. SPIE*, **1071**, 153.

Janesick, J., Elliott, T., & Pool, F., 1988, IEEE Nuclear Science Symposium, Orlando, FL.

Janesick, J., Hynecek, J., & Blouke, M., 1981, *Proc. Soc. Photo-Opt. Instr. Eng.*, SPIE, **290**, 165.

Jorden, P. R., Deltorn, J.-M., & Oates, A. P., 1993, *Gemini*, Issue 41, p. 1.

Jorden, P. R., Deltorn, J.-M., & Oates, A. P., 1994, in *Instrumentation for Astronomy VIII, Proc. SPIE*, **2198**, p. 57.

Joyce, R., 1992, in *Astronomical CCD Observing and Reduction Techniques*, ASP Conference Series Vol. 23, ed. S. Howell, p. 258.

Kells, W., et al., 1998, *Publ. Astron. Soc. Pac.*, **110**, 1487.

King, I., 1971, *Publ. Astron. Soc. Pac.*, **83**, 199.

King, I., 1983, *Publ. Astron. Soc. Pac.*, **95**, 163.

Kreidl, T. J., 1993, in *Stellar Photometry – Current Techniques and Future Developments*, eds. C. J. Butler & I. Elliott, IAU Colloquium 136, Cambridge University Press, p. 311.

Lasker, B. M., et al., 1990, *Astron. J.*, **99**, 2019.

Leach, R., 1995, in *New Developments in Array Technology and Applications*, eds. A. G. D. Philip, K. A. Janes, & A. R. Upgren, Kluwer, p. 49.

Lessor, M., 1990, in *CCDs in Astronomy*, ASP Conference Series Vol. 8, ed. G. H. Jacoby, p. 65.

Lessor, M., 1994, in *Instrumentation in Astronomy VIII*, eds. D. L. Crawford & E. R. Craine, *Proc. SPIE*, **2198**, 782.

Longair, M. S., 1997, *High Energy Astrophysics*, Vol. 1, Cambridge University Press.

Lucy, L. B., 1975, *Astron. J.*, **79**, 745.

MacConnell, D., 1995, in *The Future Utilisation of Schmidt Telescopes*, eds. J. Chapman, R. Cannon, S. Harrison, & B. Hidayat, ASP Conference Series Vol. 84, p. 323.

Mackay, C., 1986, *Annu. Rev. Astron. Astrophys.*, **24**, 255.

Martinez, P. & Klotz, A., 1998, *A Practical Guide to CCD Astronomy*, Cambridge University Press.

Massey, P., 1990, *NOAO Newsletter*, March, p. 16.

Massey, P. & Davis, L. E., 1992, *A Users Guide to Stellar CCD Photometry with IRAF*, NOAO publication.

Massey, P. & Jacoby, G., 1992, in *Astronomical CCD Observing and Reduction Techniques*, ASP Conference Series Vol. 23, ed. S. Howell, p. 240.

Massey, P., Strobel, K., Barnes, J., & Anderson, E., 1988, *Astrophys. J.*, **328**, 315.

Massey, P., Valdes, F., & Barnes, J., 1992, *A Users Guide to Reducing Slit Spectra with IRAF*, NOAO IRAF users guides, NOAO Publications.

McGrath, R. D., 1981, *IEEE Transactions on Nuclear Science*, NS-28, No. 6.

McGraw, J. T., Angel, R., & Sargent, T., 1980, in *Applications of Digital Image Processing to Astronomy*, ed. D. A. Eliot, *Proc. SPIE*, pp. 20–28.

McLean, I. S., 1997a, *Electronic Imaging in Astronomy*, Wiley, Chapter 5.

McLean, I. S., 1997b, *Electronic Imaging in Astronomy*, Wiley, Chapter 7, p. 167.

McLean, I. S., 1997c, *Electronic Imaging in Astronomy*, Wiley, Chapter 12.

Merline, W. & Howell, S. B., 1995, *Expt. Astron.*, **6**, 163.

Miller, J. S., Robinson, L. B., & Goodrich, R. W., 1988, in *Instrumentation for Ground-Based Optical Astronomy*, ed. L. B. Robinson, Springer-Verlag, p. 157.

Miyazaki, S., Sekiguchi, M., Imi, K., Okada, N., Nakata, F., & Komiyama, Y., in *Optical Astronomical Instrumentation*, ed. S. D'Odorico, *Proc. SPIE*, 1998, **3355**, 363.

Monet, D. G., 1992, in *Astronomical CCD Observing and Reduction Techniques*, ASP Conference Series Vol. 23, ed. S. Howell, p. 221.

Monet, D. G. & Dahn, C. C., 1983, *Astron. J.*, **88**, 1489.

Monet, D. G., Dahn, C. C., Vrba, F. J., Harris, H., Pier, J., Luginbuhl, C., & Ables, H., 1991, *Astron. J.*, **103**, 638.

Mortara, L. & Fowler, A., 1981, in *Solid State Imagers for Astronomy, Proc. SPIE*, **290**, 28.

Neely, W. A. & Janesick, J., 1993, *Publ. Astron. Soc. Pac.*, **105**, 1330.

Newberry, M., 1991, *Publ. Astron. Soc. Pac.*, **103**, 122.

Oke, J. B., 1988, in *Instrumentation for Ground-Based Optical Astronomy*, ed. L. B. Robinson, Springer-Verlag, p. 172.

Oke, J. B. & Gunn, J., 1983, *Astrophys. J.*, **266**, 713.

Opal, C., 1988, in *Second Workshop on Improvements to Photometry*, ed. W. Borucki, NASA Conf. Publ. 10015, p. 179.

Peacock, A., et al., 1996, *Nature*, **381**, 135.

Pecker, J.-C., 1970, *Space Observations*, Reidel.

Penny, A. & Dickens, R., 1986, *Mon. Nat. R. Astron. Soc.*, **220**, 845.

Perryman, M. A. C., Peacock, A., Rando, N., van Dordrecht, A., Videler, P., & Foden, C. L., 1994, in *Frontiers of Space and Ground-Based Astronomy*, eds. W. Wamsteker et al., Kluwer, p. 537.

Pogge, R. W., 1992, in *Astronomical CCD Observing and Reduction Techniques*, ASP Conference Series Vol. 23, ed. S. Howell, p. 195.

Queloz, D., 1995, in *New Developments in Array Technology and Applications*, eds. A. G. D. Philip, K. A. Janes, & A. R. Upgren, Kluwer, p. 221.

Reicke, G., 1994, *Detection of Light from the Submillimeter to the Ultraviolet*, Cambridge University Press.

Robinson, L. B. (ed.), 1988a, *Instrumentation for Ground-Based Optical Astronomy*, Springer-Verlag.

Robinson, L. B. (ed.), 1988b, *Instrumentation for Ground-Based Optical Astronomy*, Springer-Verlag, pp. 1–304.

Robinson, L. B. (ed.), 1988c, *Instrumentation for Ground-Based Optical Astronomy*, Springer-Verlag, pp. 189–304.

Roesler, F. L., et al., 1982, *Astrophys. J.*, **259**, 900.

Rutten, R., Dhillon, V., & Horne, K., 1992, *Gemini*, Issue 38, p. 22.

Sabby, C., Coppi, P., & Oemler, A., 1998, *Publ. Astron. Soc. Pac.*, **110**, 1067.

Schaeffer, A., Varian, R., Cover, J., Janesick, J., & Bredthauer, R., 1990, in *CCDs in Astronomy*, ASP Conference Series Vol. 8, ed. G. H. Jacoby, p. 76.

Schectman, P. & Hiltner, W., 1976, *Publ. Astron. Soc. Pac.*, **88**, 960.

Schempp, W. V., 1990, in *CCDs in Astronomy*, ASP Conference Series
   Vol. 8, ed. G. H. Jacoby, p. 111.
Schmidt, M. & Gunn, J., 1986, *Astrophys. J.*, **310**, 518.
Schmidt, M., Schneider, D., & Gunn, J., 1986, *Astrophys. J.*, **306**, 411.
Smith, M., 1990a, in *CCDs in Astronomy II: New Methods and Applications
   of CCD Technology*, eds. A. G. D. Philip, D. Hayes, & S. Adelman, L.
   Davis Press, p. 31.
Smith, R. M., 1990b, in *CCDs in Astronomy*, ASP Conference Series Vol. 8,
   ed. G. H. Jacoby, p. 153.
Srour, J., Hartmann, R., & Kitazaki, K., 1986, *IEEE Transactions on
   Nuclear Science*, NS-33, No. 6.
Sterken, C., 1995, in *New Developments in Array Technology and
   Applications*, eds. A. G. D. Philip, K. A. Janes, & A. R. Upgren,
   Kluwer, p. 131.
Stetson, P., 1998, *Publ. Astron. Soc. Pac.*, **110**, 1448.
Stetson, P. B., 1987, *Publ. Astron. Soc. Pac.*, **99**, 191.
Stetson, P. B., 1992, *Publ. Astron. Soc. Pac.*, **102**, 932.
Stetson, P. B., Davis, L. E., & Crabtree, D. R., 1990, in *CCDs in
   Astronomy*, ASP Conference Series Vol. 8, ed. G. H. Jacoby, p. 289.
Stone, R. C., 1989, *Astron. J.*, **97**, 1227.
Stover, R. J., Brown, W., Gilmore, D., & Wei, M., 1995, in *New
   Developments in Array Technology and Applications*, eds. A. G. D.
   Philip, K. A. Janes, & A. R. Upgren, Kluwer, p. 19.
Timothy, J. G., 1988, in *Instrumentation for Ground-Based Optical
   Astronomy*, ed. L. B. Robinson, p. 516.
Tobin, W., 1993, in *Stellar Photometry – Current Techniques and Future
   Developments*, eds. C. J. Butler & I. Elliott, IAU Colloquium 136,
   Cambridge University Press, p. 304.
Tonry, J., Burke, B., & Schechter, P., 1997, *Publ. Astron. Soc. Pac.*, **109**,
   1154.
Trauger, J. T., 1990, in *CCDs in Astronomy*, ASP Conference Series Vol. 8,
   ed. G. H. Jacoby, p. 217.
Tüg, H., White, N. M., & Lockwood, G. W., 1977, *Astron. Astrophys.*, **61**,
   679.
Tyson, J., 1990, in *CCDs in Astronomy*, ASP Conference Series Vol. 8, ed.
   G. H. Jacoby, p. 1.
Tyson, J. A. & Seitzer, P., 1988, *Astrophys. J.*, **335**, 552.
Vogt, S. & Penrod, G. D., 1988, in *Instrumentation for Ground-Based
   Optical Astronomy*, ed. L. B. Robinson, Springer-Verlag, p. 68.
Wagner, R. M., 1992, in *Astronomical CCD Observing and Reduction
   Techniques*, ASP Conference Series Vol. 23, ed. S. Howell, p. 160.
Walker, A., 1990, in *CCDs in Astronomy*, ASP Conference Series Vol. 8,
   ed. G. H. Jacoby, p. 319.
Walker, G., 1987, *Astronomical Observations*, Cambridge University Press.
Whitmore, B. & Heyer, I., 1998, *WFPC2 Instrument Report 97-08*, Space
   Telescope Science Institute Instrument Science Reports.
Wright, J. & Mackay, C., 1981, in *Solid State Imagers for Astronomy*,
   eds. J. Geary & D. Latham, SPIE.
Young, A. T., 1974, in *Methods of Experimental Physics*, Vol. 12, Part A,
   ed. N. Carleton, Academic Press, Chap 3.
Zaritsky, D., Shectman, S., & Bredthauer, G., 1996, *Publ. Astron. Soc. Pac.*,
   108, 104.

Index